005.1 K299c
Kelsey, Robert Bruce, 1954-
Chaos and complexity in
software

DATE DUE

July 28			

DEMCO NO. 38-298

WITHDRAWN

STAFFORD LIBRARY
COLUMBIA COLLEGE
COLUMBIA, MO 65216

Chaos and Complexity in Software

CHAOS AND COMPLEXITY IN SOFTWARE

ROBERT B. KELSEY

Nova Science Publishers, Inc.
Commack, New York

Editorial Production:	Susan Boriotti
Office Manager:	Annette Hellinger
Graphics:	Frank Grucci and Jennifer Lucas
Information Editor:	Tatiana Shohov
Book Production:	Donna Dennis, Patrick Davin, Christine Mathosian, Tammy Sauter and Lynette Van Helden
Circulation:	Maryanne Schmidt
Marketing/Sales:	Cathy DeGregory

Library of Congress Cataloging-in-Publication Data
Kelsey, Robert Bruce, 1954-
    Chaos and complexity in software: Challenging the industry and the new science/ Robert Bruce Kelsey
        p.   cm.
Includes bibliographical references.
    ISBN 1-56072-669-5
    1. Software engineering. 2. Chaotic behavior in systems. 3. Computational complexity. I. Title.
QA76.758.K46   1999   99-24598
005.1—dc21   CIP

Copyright © 1999 by Nova Science Publishers, Inc.
    6080 Jericho Turnpike, Suite 207
    Commack, New York 11725
    Tele. 516-499-3103   Fax 516-499-3146
    e-mail: Novascience@earthlink.net
    e-mail: Novascil@aol.com
    Web Site: http://www.nexusworld.com/nova

All rights reserved. No part of this book may be reproduced, stored in a retrieval system or transmitted in any form or by any means: electronic, electrostatic, magnetic, tape, mechanical photocopying, recording or otherwise without permission from the publishers.

The authors and publisher have taken care in preparation of this book, but make no expressed or implied warranty of any kind and assume no responsibility for any errors or omissions. No liability is assumed for incidental or consequential damages in connection with or arising out of information contained in this book.

This publication is designed to provide accurate and authoritative information with regard to the subject matter covered herein. It is sold with the clear understanding that the publisher is not engaged in rendering legal or any other professional services. If legal or any other expert assistance is required, the services of a competent person should be sought. FROM A DECLARATION OF PARTICIPANTS JOINTLY ADOPTED BY A COMMITTEE OF THE AMERICAN BAR ASSOCIATION AND A COMMITTEE OF PUBLISHERS.

Printed in the United States of America

To my Father.

Contents

Acknowledgments		ix
Prologue:	The Brave New World	1
Chapter 2:	Discipline, Methods, and the Challenge of the New Science	9
Chapter 3:	The New Paradigm in Science	27
Chapter 4:	Chaos in Software?	53
Chapter 5:	A Conceptual Model of Chaos in Executed Programs	85
Chapter 6:	Complexity and the Ordering Principles of Experience	103
Chapter 7:	Complex Software Systems	129
Epilogue:	The Science of Complexity, the Complexity of Understanding	151
Endnotes		159
Bibliography		169
Index		181

ACKNOWLEDGMENTS

Some of the material in this book first appeared in conference proceedings for the Seventh International Conference on Software Quality and the 53rd Annual Midwest Quality Conference, and is reprinted here with the permission of the American Society for Quality and the Midwest Quality Council. Lines from "The Song of the Happy Shepherd" reprinted with the permission of Scribner, a division of Simon & Schuster, publisher of *The Collected Works of W. B. Yeats, Vol 1: The Poems*, revised and edited by Richard Finneran (New York: Scribner, 1997). Material from *Philosophical Hermeneutics* by Hans-Georg Gadamer, edited/translated by David Linge, Copyright 1976 The Regents of the University of California, reprinted with permission from the University of California Press. Sections of Stephen Kellert's *In the Wake of Chaos*, Copyright 1993 by The University of Chicago, are included here with the permission of The University of Chicago Press. Indiana University Press granted permission to reprint a section of Edward S. Casey's *Imagining: A Phenomenological Study*, Copyright 1976 by Indiana University Press. The author also wishes to acknowledge Cambridge University Press, The University of Michigan Press, Lawrence Erlbaum Publishers, and Northwestern University Press for their kind permission to reprint at no cost material under their copyright.

PROLOGUE: THE BRAVE NEW WORLD

> We do not arrive at new conceptual possibilities on the basis of experiences, but rather we arrive at new possibilities of experiencing on the basis of new modes of experience.
> Wilhelm Szilasi, *"Experience and Truth in the Natural Sciences"*

Chaos has been with us for millennia, perhaps ever since primitive humankind first distinguished order from disorder. Early *homo sapiens* didn't need Chaos Theory to understand the turbulence of a raging river or to appreciate the difficulty of telling the next day's weather from bird behavior and the colors of the evening sky. It is only in the last several decades that Chaos has become an academic rather than a practical challenge. But as soon as Lorenz, Feigenbaum, and others began analyzing the quixotic behavior of nonlinear physical systems, they initiated the second major scientific paradigm shift of this century.

First to feel the effects of the 'new science' were classical physics and classical philosophy of science. Early on, Chaos Theory picked up the assault on the philosophical assumptions of classical physics where quantum mechanics had left off. Just as we had started to get comfortable about the obvious differences between the uncertain world of wave-particles and the predictable world of falling pianos, Chaos presented us with non-linear, dynamical systems that defied typical scientific models for both prediction and explanation. These were the cornerstones of the law-abiding universe we had come to believe in thanks to advances in physics and chemistry in the 18th and 19th centuries. Quantum physics loosened those cornerstones. Chaos yanked them free.

With the cornerstones removed, the aging edifice of scientific "Truth" began to crumble. The criteria for adequate scientific theories have changed significantly in the last two decades. Gone are phrases such as "deductive-nomological", replaced with more mercurial and multivalent terms such as "connectedness" or "fecundity" (Nagel 1979, Musser 1997, Coles 1997). Not only scientific explanation but knowledge itself is being challenged by Chaos

Theory: for some, understanding a Chaotic system is a "transcendental impossibility" (Kellert 1993).

The effects of the 'new science' are not limited to the hard sciences such as physics. Most recently, Chaos has been taken up by researchers in the 'soft sciences' of psychology, sociology, economics, etc., where it is prompting an equally radical break with tradition. Evolution, social structure, human behavior, even "mind" itself have emerged from the debris of old mechanistic and reductionist theories and are now thought of as holistic, self-referential, and chaotic "deep structures" in the world.

It may also be true that the 'new science' with its new objects of study is only the tip of the proverbial iceberg. Some Chaos and Complexity researchers have gone so far as to hail Chaos as a fundamental social paradigm shift, not just a scientific one. They take the view that Chaos regains for humankind its purpose and its role in a universe from which we have been estranged since Newton's *Principia* with its clockwork cosmos that didn't need us around to wind it or replace the battery (Goerner 1995).[1]

Clearly, Chaos and Complexity are more than an approach to solving problems in thermodynamics, meteorology, or astrophysics. They provide powerful metaphors that have stirred the imaginations of writers in and out of the physical sciences.[2] For some pioneering researchers in the humanistic disciplines of sociology and psychology, the tenets and concepts of Chaos and Complexity have provided a conceptual framework for helping us sort through the complicated behavior of people, organizations, societies, and cultures. It is certainly intriguing to ask if they can also help us understand "software" in all its senses, from software project management, to software development methods, to software product quality assurance, and perhaps even to customer satisfaction with a software product.

The success sociology and psychology have enjoyed using Chaos and Complexity certainly suggests they can help us understand software. On the other hand, the difficulties those soft sciences have experienced transposing Chaos Theory and Complexity from physical science to the human sciences suggests that such a transposition is neither simple nor without significant risks. Still, the state of the software industry today, and the serious problems facing the industry if software products (or production) are potentially Chaotic or Complex, add up to a challenge the industry must face, and must overcome.

Software is expensive to produce and notoriously unreliable, yet our dependence on software, from home banking to medical technology to governmental oversight, increases everyday. For all its tools and techniques and methods the software industry still cannot demonstrate the type of consistent and precise production we have come to expect from other industries with more credible claims to 'scientific' and/or engineering practices. Paradoxically, the industry is not yet disciplined or scientific enough to address its production processes and products from a classically deterministic point of view, and it is therefore severely handicapped when it comes to exploring the Chaotic or Complex features of software production and software products. To understand "software" in all of the ramifications listed above, the software industry will need help from sociologists, psychologists, philosophers, linguists, etc.

We should not however expect that everything we learn about Chaos and Complexity will help us manage projects or create better products. Nor should we expect to just appropriate wholesale anything the natural sciences have "done" with the new science. After all, Chaos Theory and Complexity in physics describe how things are, not how they should be. The software industry works within a value-laden universe of likes and dislikes; physics *is*, like it or not. "Cool" is a reason to use a software product. In physics, it's merely an accidental "quality" humans ascribe to a thermodynamic state.

Further, there is an historical tension between the physical and the human sciences. The laws and logic of the one do not seem to apply completely or comfortably to the other. The history of science, philosophy, and psychology is strewn with attempts to show why the "chemistry" of human relationships is, or is not, just like the "chemistry" of acids applied to metals. Software, whether we see that as its development or its use, is a decidedly intentional and difficult human endeavor. Acrid as discussions about process improvement may get, the role of the change agent can never be described with the same accuracy and specificity as the role of reagents in a chemical reaction.

And as we will see, there are indeed some significant differences between physical objects and the executing software program. These differences are, to use the jargon of the new sciences, "fecund". They suggest new ways of looking at software production, software execution, and software quality assurance. They open up new possibilities for research and experimentation that lead us away from techniques and tools to structures and emergent behavior. To take advantage of these possibilities, we must first transpose the concepts and tools of Chaos and Complexity into the environment of the software industry and then analyze software production for Chaotic behavior. Those are both daunting tasks, and in fact this book is concerned primarily with the first - transposing Chaos and Complexity from the hard sciences into the software industry. It's not simply a matter of applying a metaphor here and a concept there. Paradigm shifts are never that easy....

First, Chaos and Complexity invite us to look at software - both the product and the production "process" - as a set of structural relations rather than as a thing or a set of processes. The "pieces" we are accustomed to, such as requirements, code, and users, become limiting conditions or boundaries for "systems" such as writing code or requirements reviews or development projects as a whole. As we'll see later on, figuring out which "systems" are important and what their "structures" are is crucial to understanding how Chaos and Complexity affect software.

Second, the new science as science requires a re-evaluation of the assumptions and attitudes underlying how we think about "things" and "how things work." It is this second implication of the new science that is most problematic for software. The software industry shares with the sciences many attributes, but it lacks the one attribute it needs to accommodate Chaos and Complexity. The physical sciences and some of the social sciences have a methodological basis for evaluating theoretical applications. But software lacks such a criterion for evaluating its techniques and its methods. The industry inclines towards the uncritical acceptance of a technique or a method if it has been "successful" in some particular environment. Portability is just assumed: seldom do we probe what "successful" means for different environments. Yet this is precisely the kind of information we will need to place software development on a firm scientific and methodological foundation.

I have already referred to the new science as a paradigm shift within physics, and I suggest that Chaos and Complexity will force a paradigm shift in software development similar to the shift the biological and social sciences are currently enduring. And like those disciplines, we must learn to distinguish legitimate from illegitimate uses of Chaos and Complexity. When applied to software, terms such as "Chaotic", "nonlinear", or "dynamical" simply cannot have the same meaning they have in physical Chaos Theory. The software industry should not uncritically appropriate terms and concepts from physical

Chaos, nor should it expect its view of Chaos and Complexity to exactly match that of physics.[3]

If the industry is to understand how Chaos and Complexity affect the production and use of software, it will need to understand the methodological criteria available to us for evaluating the different approaches to software development and quality assurance. In the next chapter we will examine the issues of proof, effectiveness, accuracy, and precision in order to identify some of the conditions under which 'tools and techniques' can be considered actual methodologies. The intent is not to champion or vilify one school of thought, or method, or technique. Instead, the goal is to show where our current discipline falls short of the detailed analyses it needs, not only for defending current practices, but also for accommodating Chaos and Complexity.

And even when there appear to be good methodological reasons for accepting one practice rather than another, we will find that beneath our reasoning lie assumptions that Chaos and Complexity call into question. By the end of the next chapter, we will have examined some of these classically scientific assumptions and perspectives that imbue software development and product qualification in general. One of the paradoxical themes of this book is that we have to get more precise, more disciplined, more scientific in order to confront the new science, which presents us with an imprecisely known, dynamically changing, nondeterministic, and sometimes flat out confusing reality. The bulk of the third chapter describes in detail how Chaos and Complexity have changed that reality and how we understand it.

With chapter four, Chaos and Complexity part ways for a time. "Chaos in Software?" examines some of the problems faced when applying tenets of physical Chaos Theory to software systems. The physical sciences demand mathematically sound modeling. The social sciences (which provide the starting point for analyzing software systems) do not. But if "Chaos" is to be more than an honorific title we bestow on some system that has eluded our understanding, we have to know the attributes of Chaotic software systems and where to look for them. Drawing on the work of David Harvey and Michael Reed, we will use the notion of a "pluralistic methodology" to help us sort through the kinds of software systems available for study, the various attributes of software systems, and the rationale for using different strategies to analyze them.

To exhaustively analyze all such systems will take several years of research. Not all of them will prove to be potentially Chaotic. But in chapter 5, "A Conceptual Model of Chaos in Executed Programs", we will look at one particular type of software system - the program as executing in some physical environment - that does possess Chaotic potential. Once again we will encounter the problem of using tenets of a physical theory outside the realm of physics, and we will spend some time developing software-specific definitions for Chaos concepts such as initial conditions and turbulence. The conceptual model of Chaos presented in chapter five is certainly not complete – it focuses on but one of the many software systems identified in chapter four. But with respect to its structural components, that is, its transposition of physical Chaos to software systems, it should provide a useful starting point for future analyses and for other models.

It makes sense to address Chaos, initially at least, separate from Complexity. Research into Chaotic behavior in development projects, in defect rates, and in a number of other areas can proceed independently of any hypotheses about Complex behavior in software systems. The goals of *Chaos and Complexity in Software* will not have been met, however, until we have at least a preliminary understanding of what Complexity in software is and where it is likely to occur. The most difficult part of accommodating the new science in software is learning how to think in the style of the 'new scientists'. When it comes to understanding Complexity, understanding Chaos turns out to be merely a warm up exercise.

Complexity is not just a new tool we apply to old problems. It changes the problems altogether. If Complex systems are not just aggregates of pieces, then what are they? What are the structural components of Complex software systems? If they somehow evolve over time, then is it true that what we observe of them is "true" only within a limited chronological window? Are the analytical strategies available to us for Chaos also applicable to Complexity? Are there special limitations imposed on our ability to understand a Complex software system?

These kinds of question are difficult to answer even in the hard sciences, and they become more difficult for the software industry. Complex physical systems can be modeled without getting involved in sentience, intentions, goals, and a host of other confusing and elusive notions that the hard sciences don't well equip us to study. But we cannot avoid such issues if we want to study software systems, some of which share more characteristics with social systems than with physical ones.

Much of the existing research in the social sciences approaches Complexity in social systems as a special form of Complexity in physical systems. In the sixth chapter, "Complexity and the Ordering Principles of Experience", I introduce an alternative view. Software is an artifact of human understanding, encoded in programming language. Its linguistic component and its purposiveness set it apart both from typical physical systems and from social systems treated as collections of empirically catalogued behavior patterns. Chapter 6 identifies the structural characteristics of 'lived experience' and offers some simple illustrations of how these affect the development of software product.

But not all structures produce emergent behavior. And not all emergent behavior is Complex behavior. The seventh chapter has the unenviable task of correlating time, understanding, emergence, and software Complexity. The focus there will be on the conditions for Complex behavior in software, although the "conditions" we'll pursue will be philosophical rather than programmatic. Unlike Chaos, where we can identify some product characteristics that contribute to Chaotic behavior, Complexity has more to do with understanding and experience than with particular algorithms or product types.

There are risks to an expedition like this at a time like this. Just as the software industry is beginning to recognize the need for more discipline and greater reliance upon traditional "quality tools", Chaos and Complexity arrive on the scene to challenge the validity of those tools and the value of their application. My goal in this book is to explain what software-specific Chaos and Complexity are and how they manifest themselves in software execution. Readers who believe that software production and reliability will one day be formulaic and predictable may find this analysis disturbing, perhaps even wrong headed. But the burden of proof is on the industry; this is not the time for 'religious wars' over Truth, Science, and Discipline. Only additional research will show whether software Chaos can be tamed (perhaps even avoided). In that respect, *Chaos and Complexity in Software* should be seen as an invitation not an affront to classically oriented research.

On the other hand, the view of Complexity presented here is very likely to be controversial no matter how accommodating readers are towards my views of Chaos. I suppose I could find solace in the fact that it is one of the occupational hazards of software quality managers to be branded trouble makers. But my motives for presenting Complexity here don't come from devotion to some perverse desire for infamy. Instead, I am trying to look forward to what science, not software, will be like in the years to come.

The focus in the later chapters of the book is on the experiential aspects of software development – hardly the usual fare for a 'scientific' analysis. But Chaos, Complexity, and new theories of consciousness deriving from quantum physics and evolutionary studies appear to be replacing the Subject-Object paradigm with a more 'cooperative' system. What exactly that means isn't certain, though there's no shortage of discussion. But what is certain is that the old view of "science" as quantification of attributes of objects is neither universally viable nor particularly productive in many sectors of research.

Today, many quality professionals see statistical process control as perhaps the paradigm of applying scientific method to quality assurance. In the years to come they will have to contend with definitions of scientific methodology and scientific knowledge that challenge, even threaten, the assumptions behind that view. The industry will soon need to understand the methodological limitations of "empirical" and statistical analysis, and it will no doubt have to rethink the role quality assurance can play in the brave new world of Chaos and Complexity. Looking ahead towards that eventuality, in the Epilogue I will try to place my analysis of Chaos and Complexity in the larger picture of the evolution of science and of our knowledge of ourselves.

Chapter 2:
Discipline, Methods, and the Challenge of the New Science

> The attraction of knowledge would be small if one did not have to overcome so much shame on the way.
> Friedrich Nietzsche, *Beyond Good and Evil*

Is Software Development a Discipline?

As I write this, less than 500 days remain until the year 2000. The World Wide Web job boards are full of postings for Y2K project managers, analysts, and test engineers. "Wanted: Brave, industrious individuals to participate in corporate-wide mainframe conversion effort. Experience the excitement of an archeological dig without ever leaving your desk!" Speaking of artifacts, Cobol was for years something akin to the curse of opening an Egyptian tomb, but it is once again a legitimate line item in Information Technology resumes. What goes around, comes around. If nothing else, Year 2000 efforts will force the repair or replacement of some very ugly code.

No matter what the rationales were for the designs and development practices that brought us to this impasse, the fact remains that the 'Year 2000 Problem' has tarnished further the reputation of the software industry. It was bad enough when the databases of the industry's own research organizations such as the Quality Assurance Institute, Software Engineering Institute, or Software Productivity Research were filled with dossiers on failed projects, cost overruns, and defect-ridden products. But now every airline traveler, every bank customer, and any customer of state or Federal service agencies knows that his or her credit rating, pension, entitlements, perhaps even his or her life itself, is at risk.

Is the Year 2000 problem just one more example showing that the software industry in general doesn't understand how to develop durable, reliable products? In our own defense, we can cite some of our success stories. Some organizations have demonstrated that they can produce virtually defect-free software. Some organizations have met time to market windows with passably functional products and managed to retain customers long enough to get them the fully functional version a few months later. And some organizations release product at a fraction of the cost of their competitors and somehow manage to stay in business and out of the liability courtroom. Isn't this enough? After all, the motto is "Good, fast, cheap, pick *two*," not *three*.

Obviously, it isn't good enough. Few in software development would say it is. The proliferation of standards and how-to books and the profitability of consulting firms show that there is a strong market for "continuous improvement." One can study the IEEE standards, or the Capability Maturity Model, or the ISO standards. A half dozen publishing houses carry significant lists of texts on software engineering and quality assurance. Those who prefer the learning environment of the classroom or lecture hall can choose from hundreds of courses, seminars, and dozens of conferences offered each year. There are professional organizations to provide networking, information exchange, and camaraderie for those struggling to tame the beast of software development, and professional certifications for those who dare claim some proficiency in the ring.

But for all the activity, is there significant progress? Some of the experts, and some of the evidence, suggest there isn't. This or that standard has been openly criticized in print as inadequate (Bollinger 1991, Krensky 1995, Paulk 1995 App. F, Zuckerman 1996, QAD 1998, Wang 1997b). Professional certifications and university degrees notwithstanding, whether there is *a discipline* of software development is subject to debate. While some would elevate software approaches to the status of theories (Shaw 1990) and software to an adolescent discipline (Parnas 1998, Ebert 1997), others deny approaches that honor (Beizer 1995b) and wonder if software is a professional discipline at all (Kaner 1997). We can't look to the academic community to help resolve that dispute. One can scan the course listings of colleges and universities or the job postings of *The Chronicle of Higher Education* for a long time before finding references to quality assurance methods and standards (Weinstein 1998, Evans, J. R. 1996).

And down in the trenches, there are so many methodological choices one needs not one but several scorecards to keep them straight. In quality systems we have ISO 9000/9000-3, the Capability Maturity Model, SPICE/ISO 15504, and various other models (e.g., Wang 1997a). Implementing just the ISO 9000 standards themselves has been the subject of several treatises: Jenner 1995, Oskarsson 1996, Sanders 1994, Schmauch 1994, Radice 1995. Interested in engineering models? Try Pressman 1997, Kan 1995, Humphrey 1989, McClure 1981, Rumbaugh 1991, Drabick 1997a & 1997b, and QAI's *Solutions* manual, just to name a few. The literature on metrics is enormous, from classical texts such as Grady 1987 & 1992 to more advanced and specific approaches such as Poulin 1997. When we tire of writing test plans, we can turn to the works by William Perry, Boris Beizer, Cem Kaner, Brian Marick, J. H. Dobbins and others for some diversion. This list goes on and it is growing every day.

So it should not be surprising that if one were to visit a hundred development shops, from Department of Defense contractors to large and small MIS groups, one would find different ways of doing just about everything. If one then asked why there doesn't seem to be much consistency in the way software development is conducted, one might hear one or more fairly standard replies. "Any approach that has survived peer review for conferences or journals, or has withstood the pressures of the competitive consulting marketplace, can boast some success." "The 'right' approaches are out there, but not all development organizations are aware of them, let alone use them." "Computer science curricula teach languages and programming, not project-wide development methods, so many development organizations have to find their own way through the jungle."

These may be accurate descriptions of the situation, but is the situation that bleak? Is the choice of one particular approach rather than another influenced by sociological and financial concerns – where the development team lead went to college, which conference the development or quality manager recently went to, or what consulting firm the company can afford to retain?

I am not for a moment suggesting that software professionals should give up in despair and go off to become carpenters' apprentices. Many development and QA teams have made tremendous progress despite lack of management support or in spite of the absence of a uniform and coherent quality system in the organization. And many of them have also had to contend with technologies that advance so quickly one barely has time to finish the online tutorial before another tool or platform is released. But it is reasonable to probe into the reasons why progress comes at such a cost – in dollars, lost sleep, burn out, divorces, etc. – if it comes at all. What is "software" doing wrong?

To listen to the proponents of some quality methods or development methodologies, the answer is simple. Software could "get it right" if it just complied with [enter some standard here], or if it dropped the current platform and used [enter some language and platform here], or if it used [enter some test techniques here], or if it applied [enter some reliability equation here]. Even if we ignore the zealous marketing claims of different proponents, a fundamental issue remains. Why are there *so many* such approaches when each is in one or more ways normative and proscriptive? Is the software development industry perhaps comprised of competing schools of commercial art, each with its own "style" and "techniques" that it presents as *the* way to do things?

Most software professionals would instead answer that some of these approaches are "better" than others, and the software profession just hasn't weeded out the less effective or less comprehensive ones. But that answer simply begs the question: what do "better" and "effective" mean, how are they defined? Are they defined the same way for all software types, all development organizations, and all products? If not, what distinguishes "better" for one situation from "better" for another, or "effective" here as opposed to there?[4] Eclecticism *per se* is no solution because that doesn't address why this collection of approaches is better than some other collection, or why the specific approaches in the collection are legitimate. As soon as we speak of a software *discipline*, or of software quality *assurance*, or of product *warranties* then we have to address "methodological justifications" for what we do.

APPROACHES VERSUS METHODOLOGIES

On the face of it, we might appeal to our various tools, techniques, standards, etc. as evidence of discipline and methodology. But the fact that there are so many such approaches to "doing software" makes this appeal less than satisfactory. Suppose 100 people tell you 100 different ways to make something the "right" way. Each of these 100 claims is, to a greater or lesser degree, normative and proscriptive: they prescribe one process or approach rather than another and do so presumably on the assumption that the approach reproducibly provides results consistent with the logical and material constraints of the process. You are now faced with a dilemma. Either there is a profound disagreement over what "right" means (which also suggests there is disagreement over what criteria are to be used to evaluate the product), or the product is in some ways so independent of the method of production that you can get it "right" (or not) regardless of how you make it.

Neither of these interpretations is particularly palatable. The first presents us with anarchy and relativism. There is no immediately apparent reason to choose one approach rather than another. Even if the proponent of approach #72 could produce evidence that using the approach always produced the same results, the proponent of approach #15 could claim that those results were not acceptable. But one can hardly take the second interpretation seriously, either. I've never met a software professional who introduced him or herself as "a hacker". I know of no development teams that sacrifice to the God of Luck when they ship a product. (Libations and toasts don't count here - Bacchus was a better excuse than he was a deity.) "Here a miracle happens" is not a line item on project plans and on the whole it is decidedly *not* how we see software development.

In fact, the software industry is staunchly committed to a deterministic and predictable world, despite the fact that its track record often indicates that at least the world of software development is neither deterministic nor predictable. We follow the procedures and standards and norms that we do precisely because we believe they 'work'. And when they don't work we look for the reasons, the causes of the failure. Someone or something caused the product to be defective, the project schedule to slip, the design to be inadequate. Such notions of determinism and causation lie at the heart of all those normative statements, procedures, standards, methods, etc. that fill the pages of industry publications. The normative pronouncements of different "schools" of engineering practice, the elements of the different standards, and the various development life cycle models are all intended to guide development and test in directions that benefit the product, or they are intended to prevent actions that may be detrimental to the product. They all assume causal relationships between performing certain tasks and getting (or avoiding) certain results. And proponents of such approaches can all claim some degree of success and can cite these successes as evidence that their particular approach "works".

But having some evidence that this or that approach to code review or this or that design method "works" just isn't enough. Works compared to *what*? Did the development team know before hand the range of interpretations the requirements would allow, so it could determine how effective its requirements analysis and management techniques were? Did the development team know beforehand how many defects it would introduce, so it could determine the effectiveness of its defect removal techniques such as inspections and testing? Did the project manager know beforehand how much code each developer could produce in a day, of this type, for this type of product, at each stage in the project, so that schedule and productivity projections could be validated as methodologically accurate rather than as fortuitously so? How do we *know*, without an incontrovertible empirical baseline or without a scientifically or philosophically complete justification, that the techniques we used actually, deterministically, produced the effects we claim they did? We believe they did. We wouldn't use these techniques if we didn't believe they worked. That's not the issue. Can we *prove* it?

I am not convinced we can. While we have lots of data about development project performance, we seldom critically assess it for contamination, multiple contributing factors, etc. Likewise, while it is almost mandatory to include the phrase "empirical analysis" in professional journal abstracts, it is sometimes a strange kind of empiricism that's touted.[5] The software industry certainly needs empirical justifications for its techniques, but more importantly it needs to know where and when and why any particular methodological justification is, or is not appropriate. As we will see in a later chapter, the strategy used to analyze an object reveals only certain aspects of the object and any strategy has certain limitations.

In short, software professionals have an obligation to find a rationale for their efforts, to find a "methodological and epistemological foundation" to use the philosophers' terms, especially since there are multiple, sometimes opposed, approaches at their disposal.[6] Can *any* particular software development approach claim *any* legitimacy other than 'some people believe it has worked for them'? Can the choice of approach(es) be justified (in some sense) methodologically or scientifically?

SCIENCE AND SOFTWARE

There are certainly some superficial similarities between the scientific and software communities. Like chemistry, we have our table of elements and rules for creating compounds; we have code, macros, functions, operations, classes, objects, modules and builds. With biology we share a knack for taxonomy: there are fault trees, inheritance trees; hominids we divide into QAI CQAs and ASQ CSQEs. Our measures and metrics could be the envy of every research lab in the country. We have lines of code and function points, and several ways to measure reuse. We can choose from multiple taxonomies for defects found in inspections, we have at our disposal multiple quality factors for use in quality planning. We can analyze our defects by injection phase, detection phase, location, effect, or raw run rate. And like cosmology, we have our models of the Universe of Software, and can describe its evolution using 9000-3 or 15504, various development and testing maturity models, and/or life cycle models such as iterative, spiral, or waterfall.

And like the world of physics (at least before Chaos), the world in which software is developed appears to be a linear, predictable, deterministic one. Tasks and events proceed in a more or less orderly manner, and surprises come from outside the development effort or from bad planning, not because the code or the design is vital and malicious. The project proceeds from one line of the project plan to the next; code executes one line after another. Even iterative development styles are really straight lines in a space-time curved in on itself by the gravitational force of that black hole, Customer Satisfaction. We have control paths, data paths; dutifully coding to ambiguous requirements we often follow garden paths. We have flow charts, state diagrams, algorithms and transaction protocols. We routinely break "software development" up into parts - the requirements, the specifications, the code, the tests – and often analyze and manage each piece as if it were an independent entity.

And when it comes to software quality assurance, especially product testing, we are decidedly on the side of Newtonian Mechanics rather than Art.

We stimulate a known sample of the product (a code path) under known conditions (the test platform) and from the results we implicitly or explicitly make claims about the product's future behavior. We fix bugs, fully convinced that the macroscopic product behavior found in the test lab has as its direct and sole cause some number of lines of code. Then we retest the fix and assume that if the previous behavior does not occur, the fix was correct. We do load and stress testing, fully expecting the behavior of the product to emulate that gas-filled balloon subjected to pressure and temperature changes. We don't know exactly how the code is behaving, any more than we can see the molecules within the container: but if the product doesn't fail, if we don't burst the Development Manager's balloon as it were, we assume the product will behave the same way in the hands of the customer. For software quality assurance, the software product is a mechanical (though perhaps impenetrable) device. It has definable and observable states, and future behavior can be predicted based on current behavior.

With so many similarities between software and various sciences, is it possible that one day software development will be included in that "big honker binder" physicists dream of, that "Grand Unified Theory of Everything"? Will we be able to reduce software engineering to the "physics of chips and bit streams" and to reduce software development project management to the "thermodynamics of bodies under stress"? In the era of classical philosophy of science (e.g., Nagel 1979), such reductions were not only thought to be possible, they were a requirement. No discipline outside of physical science could claim any intellectual rigor without knowing its laws and the relationship between those laws and the laws of physics. That agenda is not as popular today, and Chaos and Complexity have made the assumptions behind the agenda untenable for a large class of events and systems. But even if Chaos and Complexity posed no threat to the software industry, the industry's current disarray would prevent it from claiming any scientific bases for its practices.

A scientist engaged in an experiment has good reason to believe that the experiment will be successful precisely because the experiment is based on hypotheses and/or theories. While we might like to imagine scientists as Dispassionate Objective Minds Examining the Innate Behavior of Matter, the reality is that scientists (like the rest of us) often see what they expect to see (Achinstein 1993, Feyerabend 1986, Lewenstein 1997, Pinch 1997, Kelsey 1995a). The experiment provides experiential evidence of the truth of a general proposition about the way things behave just because the experiment has been conceived and conditioned to address the hypothesis or theory.

Once an hypothesis has been substantiated by experiential data, it can be used to predict other events. Predictions assume that events can be described using deductive logic: knowing the relevant state variables and their interrelationships for system S at time T allows one to deduce the state of system S at time T+x. That assumption is based on another: there is a deterministic relationship between one set of events and another. It is this explanatory, deductive, deterministic reasoning that sets science as a body of knowledge apart from science as a collection of data points.

At first glance, software approaches might appear to be a kind of theory. We know that any approach has grown out of the experience of its proponents. They expect that an approach will work in a future implementation – it will be confirmed in that experiment – because they think they have found a general 'law' of behavior in software development. The international standards are standards (one hopes) because they describe a set of actions that have predictable effects on the development of quality software. In a way, they are 'laws' of development, albeit laws that are obliquely expressed.

But the experimental physicist is better positioned to generate a viable theory or an accurate prediction. The physicist knows (or learns through experiment) the components of the system s/he is working with, the physicist has modeled the interrelationships between components, and the physicist can express the state of the system in non-value laden terms. I suggest that software process proponents do not know (or at least do not always articulate) all the components of the software "system" they are working with, or if they do they have not modeled all their interrelationships. And if they have the model they still cannot express the future state of the software product in non-value laden terms without significantly restricting what they mean by "product". We'll explore the first and second limitations in more detail throughout this book. For the moment, let's look at a simple example, keeping in mind that I am deliberately not trying to advance one "school" over another.

Consider the normative approach "Document and test against requirements." Hardly a controversial maxim, but it *is* a normative approach. After all, there's no Act of God that says you can't produce acceptable product by testing everything *but* conformance to the requirements; you might just be enviably lucky and the developers did everything right. But our collective experience has shown us that in order to produce the product that the customer wants, we need to test the product to verify it actually meets their needs. This has proven easier to claim than to complete, so over time another procedure has been added to this approach: "Create a requirements/test matrix to use as a checklist to ensure you are addressing all the requirements." With this addition, we are trying to identify the interrelationships between two parts of the software development effort, the requirements statements and the actions used to verify that the product conforms to them.

But as many of us have learned the hard way, one can verify that the product meets the requirements, first time, every time, and still hear "But that's not what I asked for!" To limit exposure to imprecision in requirements and to requirements drift, we enhance this approach: "Ensure you have the requirements defined properly using joint customer-development reviews, and manage changes to those requirements in such a way that any proposed changes are distributed to and reviewed by all affected parties."

While I wouldn't claim this is an accurate or complete history of 'testing to requirements' in the industry, this example does illustrate that we incrementally increase our tasks and procedures, usually in response to defects or deficiencies in the currently implemented set of tasks. That's a good thing of course, fully in keeping with our goal of perfecting our tools and techniques. We can test to requirements, but we need a way to show our tests were correct with respect to the requirements. So we create a checklist. We find that the requirements may be imprecise, so we create a way to improve their precision and correspondingly improve the precision of our ability to test any set of requirements (joint reviews). It's worth reiterating for future reference that these kinds of improvement are often incremental – they often emerge from our collective experience trying to "get software right" over many years.

And this example also illustrates what I mean when I say we often do not articulate all the parts of a system or articulate the interrelationships. The approach above does not address coverage and it does not address code complexity, either of which may affect how effective "testing" is. The complexity of the product may mask many potential errors that may not be stimulated by requirements-based, function-based testing alone. The schedule plays an important role, too. A tester performing a manual script in the 11th hour of his 8th straight 14 hour day may miss an error message and inadvertently allow a bug to be released. And we haven't addressed the development platform, development methods, etc. This approach simply does not take into consideration all the components of the "system" that is product development, therefore any claims that it makes about the interrelationships between its featured components are suspect.

There's nothing "wrong" with the approach outlined in this example. It's simply incomplete. But completeness is critical if we are to understand where a process is applicable, how it is to be applied, and what interdependencies exist between this process and others. At issue is "improvement" as opposed to "change". If we devise a means to improve a particular aspect of the software production process, we should know whether the proposed change impacts, or is affected by, other aspects of the production process. Put differently, we should treat process improvements as "sustaining engineering" rather than "point fix maintenance". We should no more change our processes without concern for integration with the rest of the production process than we should be changing code without concern for our change's impacts and dependencies on the rest of the code.

One might argue that we have to start somewhere: if we waited for proof that an approach was universally applicable, even if we waited for evidence that it was effective in multiple environments, we'd never make any progress in software process improvement. Unlike the physicist (so this argument might go), software professionals have to ship product. There's no time to fully describe every facet of development and qualification, no time to generate hypotheses that we could, through experiment and verification, elevate to the status of theories. I'm not sure that research scientists competing for limited grant money and pressured to "produce" something from their research would agree that they have it any easier than software professionals do. But they do seem to be better at understanding the limitations of their hypotheses than we are at understanding the limitations of our approaches.[7]

Putting aside for the moment the case of "paradigm shifts", the sciences also advance incrementally; they too "have to start somewhere" and the scientific body of knowledge advances through trial and error and through critical review of theory and evidence. It is not so much a matter of who has the grant money and time that sets software engineering and physics apart, it is more a matter of the discipline of verification. If we "test" an approach out in an environment where there are other unknown and possibly competing forces involved, the "test" isn't a fair test. If the test is a success, we can't claim with complete certainty that it was due to the approach. If the test fails, we can't claim with complete certainty that the approach was at fault.[8] How often do we seriously apply the regimen of test methodology to our implementations of software process improvement approaches?

A viable scientific theory, on the other hand, has a scope of application, it contains or references mechanisms for applying it and through which it can be validated, and it takes into account all the characteristics of the object being theorized about (or offers good reasons why it does not have to do so). It is subject to scrutiny because it explicitly states the what, when, where, and why of its application. Its effectiveness, accuracy, and precision can be predicted and verified. Or not, as the case may be. Theories come and go; just because a theory meets the "structural" requirements of scope, application, and completeness, it is not therefore "true". But these structural characteristics do make the theory verifiable or refutable. The theory works or does not work in the proposed scope. It does or does not provide useful predictions about nature. It is or is not complete. "Scientific knowledge" is primarily criteriological not ontological; the "things" of the classical Newtonian universe have been replaced by the "probabilities" of the quantum universe, but the "how" of making credible scientific statements has not changed terribly much.

JUSTIFYING APPROACHES

We will find in a later chapter that treating software as a "thing" is but one of several strategies one can use to analyze it. And we will also find that even this "thing" is less like a podium and more like the sound system that carries the voice of the speaker across the auditorium. Trying to justify development techniques or qualification methods is problematic when what is being created, analyzed, and measured isn't a physical, non-significant entity like a steel rod. To the extent that the development process is a volitional, social phenomenon, it stands just outside the ken of what we normally think of as "empirical" analysis. And to the extent that the software "product" is a deliberately generated, linguistically mediated artifact, it combines attributes of physical objects, processes, and performances. We will return to such matters when we look at objects of study and objectives of study in a later chapter. For the moment we need only note that we may need to combine "empirical" case studies and statistical analyses with formal deductions from other disciplines if we want to show that a given approach is in fact responsible for the successes we want to attribute to it.

Still, the industry could do a better job evaluating the approaches it has available to it today, even without delving into issues of psychology, linguistics, sociology, epistemology, and theory of mind. The value of tools, techniques, and standards will forever rise and fall on the industry's stock market in response to sociological trends, but the industry should be able to distinguish between fortuitous and reproducible successes using criteria already available elsewhere. If an approach could demonstrate its effectiveness, accuracy, and precision, then that approach could legitimately claim it was superior to some approach that could not produce such evidence on its own behalf.

If an approach is effective, it does what it is supposed to do. By itself, this doesn't help us much since the approach might make good on its claim in one case and not in another. This is why the criteria also include accuracy and precision. In software, accuracy and precision are often considered part of the quality factor "correctness": the software does what it is required to do, it accurately reflects the requirements and it calculates and displays data with whatever degree of precision is required (Schulmeyer 1987, Dobbins 1987). This is not how I intend to use these terms. Instead, I want to invoke meanings they have in manufacturing quality management.

In the manufacturing environment, a properly calibrated measuring device is expected to provide consistent readings, and it is calibrated to be both accurate and precise. Suppose I have a manufacturing process that produces Widgets to a tolerance of plus or minus 5% of An Ideal Widget Weight. The instrument I use to measure the actual Widget weights at the end of the assembly line needs to be calibrated so that it is both accurate and precise. On a sample of Widgets with known weights, the instrument should consistently register weights within some predefined tolerance for that instrument. This is accuracy. And the instrument should be able to consistently reproduce its own readings. This is precision (Juran 1988, 18.61-63).

Since we are dealing with software production methods and not with measuring instruments, we need to modify these definitions slightly. We need to take into consideration the fact that we are dealing with ways of generating software products, not with measuring instruments *per se*. And we need to take into account the fact that software development organizations build unique products. Once the CD-ROM ships, that development effort is over, that "product" is finished. There's no "going back" to build the *same* product all over again, and any subsequent development effort, even if it is a maintenance release for the "same" product, is really an entirely new development effort. With those caveats in mind, we can define accuracy and precision this way. Approaches used to generate and test software are accurate if they can produce results within a defined tolerance level. And they are precise if they can reproduce their results on similar efforts. Figure 2.1 offers a simple illustration of these concepts.

Figure 2.1 - Accuracy and precision in methods or approaches.

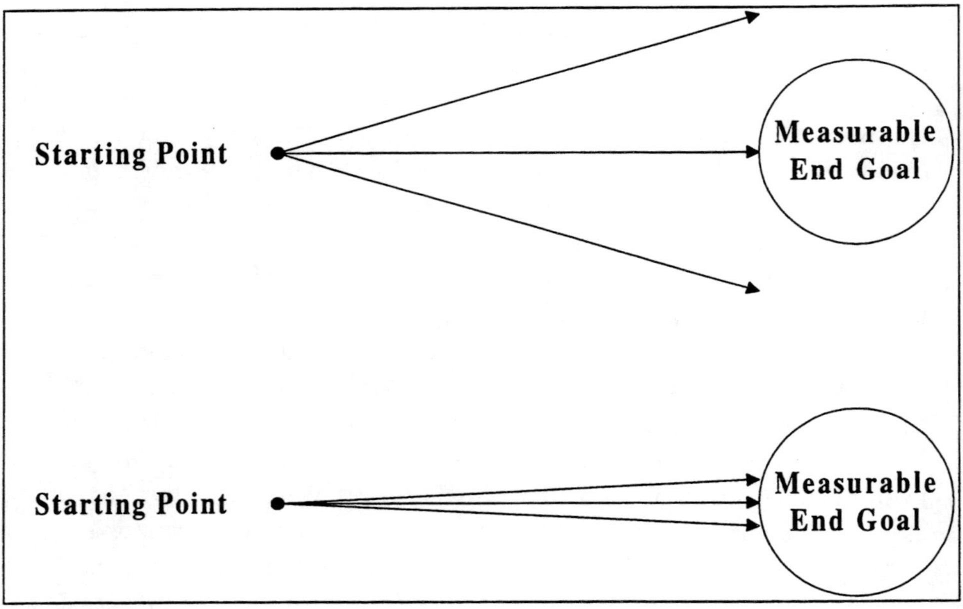

The top approach is neither accurate nor precise after three runs. The bottom method is both accurate and precise after three runs.

So if writing all the code backwards from the final line to the first line "works for you," if it's effective, that's a single instance of what *may* be a viable approach. That way of writing code still has to show similar effectiveness for other development efforts for similar products, and of course "effective" has to be definable, measurable, and within some tolerance consistent across the development efforts. If writing code backwards can do that, it is a viable approach. A single observation of a physical system is not grounds to develop a scientific theory of that system's operation, and a single successful release does not a method make.[9]

I anticipate that proponents of the better known, more comprehensive, and more widely adopted approaches are probably saying to themselves "We've known this all along and we can meet the criteria of effectiveness, accuracy, and precision." That may well be. But my intent thus far has been to show that we can't accept such approaches just because they are popular or widely adopted, any more than we can reject the Write Code Backward approach because it is *not* popular or widely adopted. I have been interested in finding criteria that are independent of sociological or historical or academic pressures, criteria that force the disclosure of evidence, criteria that will indicate limitations on applicability. If such criteria happen to elevate existing 'schools' such as UML or Cleanroom to the status of viable approaches, all the better. That way, at least, the "discipline" of software won't have to start from scratch.

THE CHALLENGE OF CHAOS AND COMPLEXITY IN SOFTWARE

I have suggested that we shouldn't place too much stock in an approach's success until we know all the factors involved in the development efforts in which it has been tested. Surprisingly, for all the models, metrics, components, and concepts listed in the opening section (and that list is by no means exhaustive), we don't seem to be able to identify all the interrelationships and interdependencies in software development. Without that, we won't be able to legitimately evaluate approaches.

Once we know the assumptions and limitations of the current set of techniques and norms and procedures, and once we supplement them with additional information from sociology, psychology, and linguistics, we will certainly find that there are several ways to "get software right". In arguing for evaluative criteria and for disclosure of assumptions and limitations, I have been emphasizing methodology over method. And, ironically, I have been preparing us to face the greatest scientific challenge to software's currently somewhat flirtatious relationship with classical science: Chaos and Complexity.

Chaos Theory shows that there are certain classes of systems that defy the classical scientific view of a predictable, static, linear, and deterministic world. And Complexity shows some of these systems are emergent – some systems in the physical world are not static and perhaps even the laws that govern those systems are not static. Later chapters will explore the possibility that some types of software can display Chaotic behavior and may show the emergent aspects of Complex systems. In those cases, our current linear, deterministic, and mechanical views of software (both its production and the product itself) are inappropriate.

If we don't understand the relationships and potentials in the complex and confusing, possibly Chaotic and Complex, environment of software development, then we can never know if an approach works or not. "This approach worked before so it will work again" becomes a vacuous claim in the face of Chaos unless the approach recognizes its own limitations with respect to Chaos. "This development model produces software to 3 Sigma" is a vacuous claim if the next product developed with this model exhibits emergent characteristics and the model does not accommodate Complexity in systems. Until the industry learns where and when Chaotic or Complex behavior can occur in software development and execution, all attempts to justify industry methods and standards remain in jeopardy.

In sum:

We need to apply scientific regimen to software process improvement approaches in order to avoid the chaos we are currently in. This scientific regimen may never attain the discipline that characterizes classical physics, but minimally it must prescribe methodological criteria for selecting and evaluating 'approaches' to software development and qualification.

We can not meet the first challenge until we understand to what extent software production and qualification is possibly nondeterministic and nonlinear, that is to say, where and why Chaos and Complexity appear in both production and product. And to understand their genesis and morphology, we will need to identify the structural components and conditions, peculiar to software production and execution, that enable and support nonlinear, dynamical, even emergent behavior.

The first of these tasks I shall leave to those more classically inclined, such as Capers Jones who is on record saying that software development should be more like manufacturing. This book is devoted to pursuing the second task.

CHAPTER 3:
THE NEW PARADIGM IN SCIENCE

> ...this most excellent canopy, ... this brave o'erhanging firmament, this majestical roof fretted with golden fire – why, it appeareth nothing to me but a foul and pestilent congregation of vapours.
> William Shakespeare, *Hamlet*, II, ii

You must forgive Hamlet's rather morose mood. His mother has just married his father's murderer, he's about to have a contract put out on him, and he's suffering from depression the likes of which the world would not see again for almost 400 years, when corporations began investigating their Year 2000 problems. His emotional condition notwithstanding, Hamlet's observation wasn't far off. Astronomers of his era were still laboring under the influence of the Mediaeval conceptions of the universe as concentric celestial orbs around the Earth. Their theoretical views influenced the "observational language" as it is sometimes called: hence the night sky was a 'canopy' or 'roof' because the spheres were, in some sense, physical boundaries. That was soon to change.

In 1656, Christian Huygens observed through his telescope three stars in the void-like darkness of a spot in the constellation Orion. He described them as "an opening in the sky" as if perhaps they were light shining from beyond the boundary of the physical universe (Jones, K. G. 1991). Fifty-eight years later, William Derham suggested that some stars and nebulae were in fact openings in the universe through which shown the Empyrean, the realm of God and the angels according to the Mediaevals (Whitrow 1970).

Telescopes of the period were awkward and imprecise compared to the equipment available to any amateur astronomer today. But they were sturdy enough to pierce through the haze of old theoretical preconceptions – though obviously observers like Derham tried to accommodate new observations into the old theories. We know now that Huygens was looking at the Orion Nebula, a vast expanse of Hamlet's "vapours" some 1600 light years away and churning in the radiation from fledgling stars like those Huygens saw. And we know now that Derham was wrong – some nebulae are actually galaxies glowing in the light of thousands of suns, not in the light of the Empyrean. But it would take more than 200 years before we could so smugly dismiss their interpretations.

Yet once we had achieved that knowledge, it would be a mere half century before we were peering back almost to the beginning of the universe and piecing together its history. Science has advanced so quickly in this century that we sometimes forget how hard won many scientific breakthroughs are. And we sometimes forget how often scientific knowledge grows not through incremental research but through revolutions in how we think about the natural world. If we need proof, we have only to look back to the beginning of this century. We entered the 1900's steeped in the classical physics of Issac Newton. By the time Bohr, Heisenberg, and Schrödinger were finished dissecting Newton's apple into its particles and waveforms, we had a new, if rather puzzling, view of the world: quantum mechanics.

While quantum physicists were busy solving problems in spacetime, matter and gravitation, scientists in other disciplines were pondering less lofty but very perplexing problems. Dissenting voices appeared in the scientific literature in thermodynamics, fluid mechanics, population biology, and meteorology: neither classical nor quantum physics could explain the behavior of some kinds of complex systems. Now known as Chaos Theory or Complexity Theory, these initial investigations into nonlinear, dynamical systems ushered in the second major scientific paradigm shift of this century.

WHAT IS A PARADIGM SHIFT?

"Paradigm shift" is used to refer to everything from a slight change of perspective to a major remodeling of our cognitive living rooms. In the broadest sense, one could call the Total Quality Management movement a "paradigm shift" because it changed how businesses viewed their workers, their customers, and their corporate goals. TQM did not bring a radical change in how we saw the natural world, humankind, and the relationship between the two. It changed how we valued parts of the world we already understood. On the other hand, the change from Newton's concept of matter to Bohr's model of the atom brought a whole new vista into view. The trajectories traced in the vapours of the cloud chamber perforated the staid, static, and predictable macroscopic world of classical physics just as surely as the long, gangly telescopes of the 17th and 18th centuries had penetrated beyond the canopy of the bounded night sky.

To make the notion of paradigm shift less abstract, try the following experiment. Figure 3.1 shows an object comprised of three circles and some connected lines. Perhaps it's a fork without a handle. Look at it quickly. Stare hard and long at it. Focus your vision to the left side of the diagram and then to the right. How many tines do you "see"?

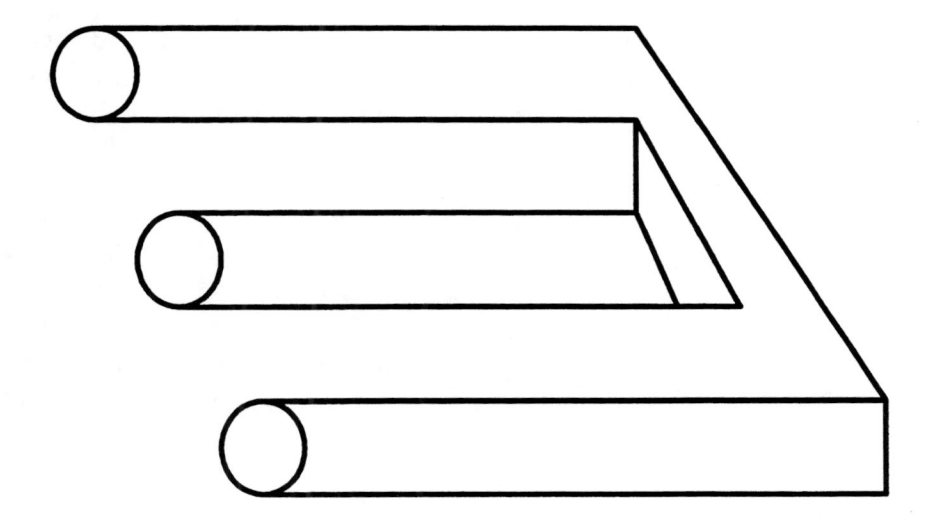

Figure 3.1 – A visual analogue to the paradigm shift.
One can perceive two different patterns from this single set of lines. New paradigms often come from the same 'data' as the old paradigm, but they prompt us to "see" or understand the data differently.

You will probably discover that you can cognitively understand that there are two views to be had here. But you cannot see and perceive both views simultaneously nor can you display in your imagination both views simultaneously. You may also find that focusing affects the perception: many viewers find that if they focus on the circles the three tines appear, but that effect is muted by focusing on the right side of the figure.

This illustrates how a paradigm shift changes the way we view and understand the world before us. The same object, the same data (here, sensory data), is combined differently in perception and recognized as one object or the other. The 2- and 3-tined forklike objects seem to be incompatible *in the act of perception*. Even in reflection we can only say, "we saw both objects". Following a distinction made by Edward Casey (Casey 1976), we can imagine *that* we could see both views at once, but we can't *actually imagine* both views at once. We see one, but "know" there are two objects available to us.

Similarly, we "know" that we as sentient hominids are composed of the entities studied by biology and chemistry yet we also "know" (albeit in perhaps a different sense) that the "self" is something more than a collection of lipids and lymph nodes, themselves composed of elements blasted out of the bowels of some dying star billions of years ago. If this sounds like we are approaching some age-old tensions – fact vs. belief, science vs. religion, empiricism vs. idealism – you're right. And, not to worry, we won't go there now – although on the fringes of quantum mechanics and Chaos and Complexity one can't escape such issues (e.g., Goerner 1995, Penrose 1996, and Hawkins 1997). We, on the other hand, need only to come away from this little exercise with two lessons in mind. First, we need to remember how fuzzy a word "know" can be – even when we think we're dealing with objects. And second, we need to appreciate that in some cases 'reality' is a cooperative effort between things, structures, and observers.

You've also just had your first brush with Chaos and Complexity. As you oscillated between the 2-tine and 3-tine perceptions, you experienced indeterminacy, the relationship between structure and meaning, and the dynamics of some systems. To use the language of the new science: the "figure" (the significance of ink stains on a page) emerged in the dynamics of your experience of it. From the same set of straight and curved lines you supplied a structure that "brought forth" the object from the book in your hand. The "object" is a structure, not a thing *per se*, and this structure is "emergent."

CHAOS AS PARADIGM SHIFT

Chaos Theory is a form of dynamical systems theory, originating in studies in meteorology, thermodynamics, and population biology. In its "purest" form, Chaos Theory is a technique for mathematically modeling and explaining bounded, complex systems. A Chaotic system may look at first glance like any typical physical system: it can be described by state variables whose interrelationship appears to be classically causal and linear. In other words, a Chaotic system is a system of interrelated components: it is not an unknowable, messy, incomprehensible collection of disparate components.[10]

But when particular state variables change in a Chaotic system, the effect of those changes is not linear. Nor is it deterministic in the same way that, say, acceleration under known force in a known environment is. Even probabilistic prediction may be of limited use: the system may transition into new structural relationships that obviate the dependencies assumed in the original projections.

Turbulence is typically cited as the quintessential example of a Chaotic system. A stream flows around a rock. Upstream of the rock, the coarse characteristics of the system can be measured, and the coarse behavior of the stream can be modeled. But after the stream passes around the rock, eddies appear; the trajectory of rivulets of the stream is no longer linearly downstream as the river turns in upon itself. Increase the speed of the stream flowing around the rock and the eddies collide, combine, or disappear. New patterns are formed. If you could divide the stream into segments for the sake of illustration, each segment would now be subject to forces from all directions, not just the downstream force of the river's flow. This proliferation of motions makes the river's flow extremely difficult if not impossible to model.

What makes this chaotic system a system? Why all the fuss about the system being indeterminate? After all, one can't predict the exact location of water molecule #28738 any better upstream than downstream. What's true of the parts should be true of the whole, so wouldn't any collection of indeterminate entities be indeterminate?

The answer, such as it is, lies in the structural relationships within the system itself. "Microlevel" uncertainties do not demand "macrolevel"

Before collisions

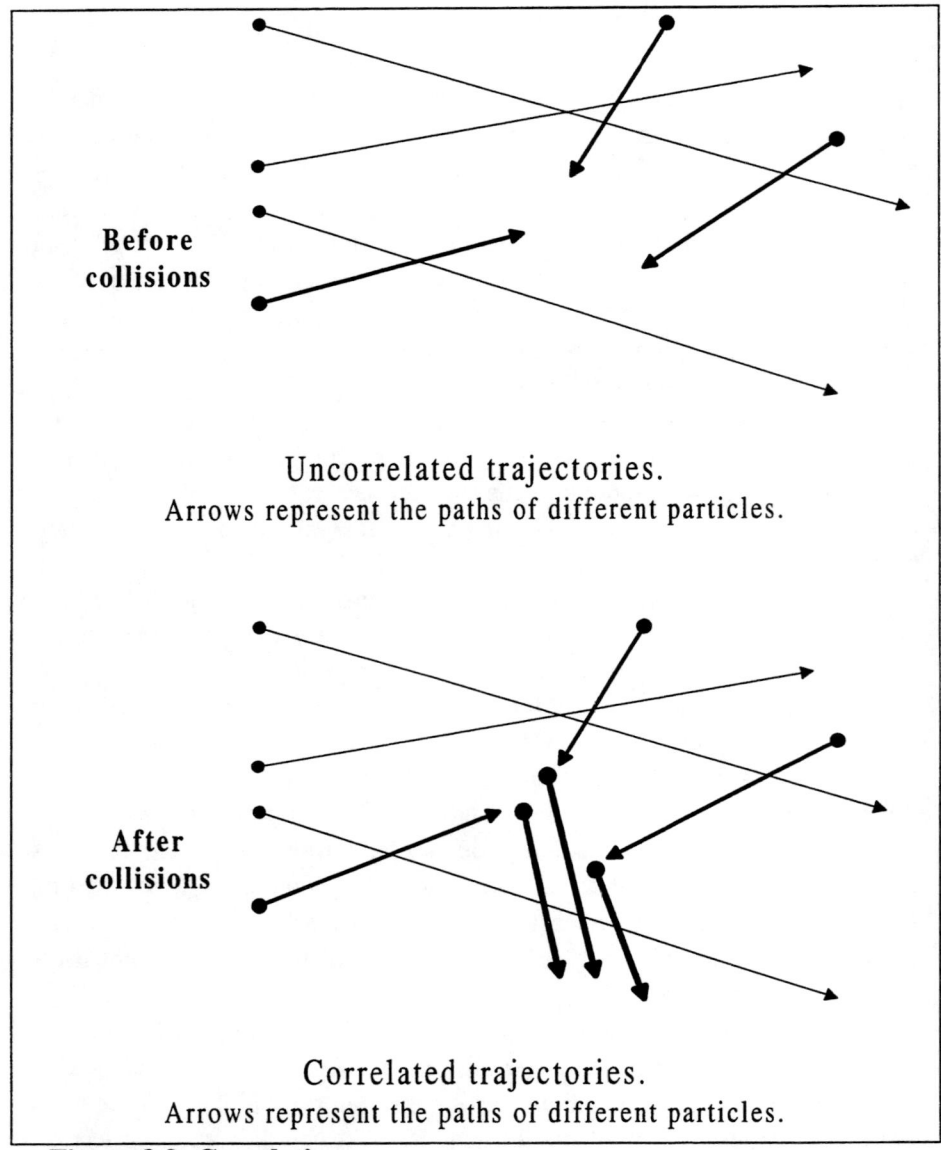

Uncorrelated trajectories.
Arrows represent the paths of different particles.

After collisions

Correlated trajectories.
Arrows represent the paths of different particles.

Figure 3.2: Correlations.
Collisions entangle the trajectories of particles. After a collision, the particle's new direction and velocity represents the 'information' it obtained during the collision.

uncertainties: regardless of where #28738 is, the river still is a recognizable entity and is subject to analysis and modeling. The motion of one molecule is not independent of the others. Collisions cause "correlations" in which molecules move together, at least for a time (Prigogine 1996; see figure 3.2). They may travel on a vector different from the downstream vector but the collective motion provides structure to the new flow. Subsequent collisions enlist more molecules. Eddies collide and cannibalize each other.

Even if we could segment the flow somehow to make it mathematically more manageable, any model of the river's motion would still have to contend with an impractical number of variables. On that view, a completely accurate and precise understanding of the river would be a practical impossibility. The notion of correlations at first glance seems to solve the problem: perhaps we can model the onset of turbulence as an increasing number of conflicting correlations? But that simply reproduces the problem at a 'higher level' of organization: a potentially infinite set of correlations still requires an impossible mathematical model (Gleick 1987). So what gives the river its turbulent coherence, why is turbulence turbulence and not complete dissipation?

That question is far from solved, although theories of the relationships between parts and wholes, between pieces and the puzzles they create, abound. Exactly how we get from particle physics to deep pan pizza, or from the probabilistic and dynamic world of quantum mechanics to the determinate and predictable world of toaster ovens and taxes is not (fortunately) essential for this book. For our purposes we need only accept that there is a chasm (albeit potentially bridgeable) between microscopic and macroscopic behavior, and that characteristics of the former affect the latter. We can be content to add one more perspective in the debate raging across the nation on automobile bumper stickers: "Things Happen."[11]

And they do so according to laws, although those laws sometimes elude us. And even when they do not escape our understanding, they sometimes don't "look" like the laws we encountered in Physics 101. The laws governing Chaos are often geometrically portrayed in phase space, wherein the state of a system taken as a whole is plotted as a single point. So plotted, Chaotic systems show a kind of periodicity or similarity of behavior over time. They revolve around attractors, locations in phase space or behavioral modes in the real world, that constrain their motions and modifications. This provides the oft-cited "order" in Chaos. It makes systems, in a coarse sense, predictable. Traffic patterns can be analyzed and predicted, but the location, speed, etc. of individual cars within those patterns may be indeterminate.

A Chaotic system has a variety of states affected by the initial values of its state descriptors. This is called "sensitive dependence on initial conditions". Changes in these initial values do not result in linear, predictable changes in the system "downstream". Instead, system behavior aggregates over time around fixed point or periodic or Chaotic attractors. When descriptor values change appropriately and the system moves far beyond its equilibrium state, the system "bifurcates". It changes its basic behavior patterns and other attractors appear (or existing attractors disappear). Chaotic systems move between stability and instability; the demarcation line is the bifurcation point between one state and another, between one set of stable structural relationships and another. Transitions happen when a system is far from equilibrium: the dynamics of the converging trajectories of its "pieces" force it to adapt to the new forces and to restructure its behavior (Abraham 1995).

The effects of Chaos on the scientific community have been well chronicled by Gleick and others, and those effects are still felt today. Complexity, which builds upon Chaos, has even greater implications for our classical notions of space, time, evolution, and knowledge, as we will soon see. For the present, let's examine three implications of Chaos Theory: the impact of Chaos Theory on classical views of scientific knowledge, the difficulty in defining determinism in light of Chaos Theory, and the need to adjust the method of study to the object of study in a post-Chaos world.[12]

IMPLICATION 1: LIMITS TO CLASSICAL SCIENTIFIC UNDERSTANDING

First, Chaos Theory makes it difficult (if not impossible) to meet the goal of detailed, accurate, and precise description that has guided the physical sciences for several centuries. At the risk of oversimplification, we can say that scientific method makes the success of a model or equation dependent on its ability to predict correctly the state of a system at a future time or to retrodict its state at a previous time.[13] That can be accomplished only if several other conditions are met. All the state descriptors must be identified (or at least the relevant ones). The effects of changes in value for one descriptor must have deterministic effects in the future – a linear trajectory, as it is sometimes called. And the relationships between descriptors must be accessible to scientific analysis and be the only relationships that matter. That is, the system is just a sum of its parts, no more; systemic laws are simply aggregates of the laws governing the pieces. These conditions are generally not met in Chaotic systems. But Chaos is more than a challenge for the scientific method; it challenges the assumptions about reality that underlie the methods.

In his philosophical investigation of the implications of Chaos, Stephen Kellert distinguishes three different perspectives in classical scientific understanding: the ontic, the epistemic, and the modal. On the ontic view, science reveals the "hidden causal processes" behind events. The ontic view is the oldest of the three perspectives, and outside of the scientific community it is the most common. This perspective informs such popular claims as "overexposure to the sun causes cancer" or "vitamin C cures colds". And it appears in quality assurance activities such as Failure Mode and Effects Analysis in manufacturing or retesting bug fixes in software where we assume a direct causal connection between the error in the code and a functional error.

In the epistemic approach, science explains phenomena; that is, it finds the logical structures and the physical prerequisites that make phenomena "predictable". (Kellert himself follows a distinction that we don't need to pursue here between an event that was to be expected as opposed to an event that can be predicted.) In popular culture, we find the epistemic perspective in attempts to explain specific behavior based on demographic attributes of the parties involved, whether that be intelligence or disposition to types of crime or to genetic defects. In software quality assurance it drives the various reliability prediction models in the software engineering literature.

And on the modal view of science, the goal is to show that events occur out of necessity rather than chance. No other conditions but the ones identified as necessary can produce the event. In software development, this perspective operates as an assumption underlying "regression" testing: if the fixed product doesn't display the same problem, then development "must have" fixed the problem.

But Chaos Theory thwarts these goals. Says Kellert:

> "...chaos theory does not provide predictions of quantitative detail but of qualitative features; it does not reveal hidden causal processes but displays geometric mechanisms; and it does not yield law-like necessity but reveals patterns." (Kellert 1993, 96)

And again:

> "Nomic necessity requires that from universal laws and statements of initial conditions we can generate with deductive rigor the uniquely determined past and future behavior of a system in fine detail. But chaos theory is neither strictly deductive, nor quantitatively predictive, nor globally deterministic." (Kellert 1993, 111)

IMPLICATION 2: THE MEANING OF "DETERMINISTIC"

Second (and perhaps as a result of the first implication), Chaos Theory has spawned much debate over what scientists mean when they say something is deterministic. As Kellert's analysis shows, the original sense – where an event occurs with logical necessity due to precursor events and in compliance with known (or knowable) laws – is clearly no longer valid for all systems. Yet there is still some debate about whether a Chaotic system is deterministic in some sense. The debate, as of this writing, often takes the form of whether Chaotic systems can be controlled (e.g., Ott 1995, Freedman 1992, DeGreene 1997, Shinbrot 1993). If one knows the general behavior patterns of a Chaotic system, and one knows which initial conditions to stimulate, why couldn't one control at least the coarse characteristics of a Chaotic system? Of course, the crucial phrase is "coarse characteristics." If one tries to control the population growth of a particular species, but one's modifications of the environment are such that the species dies off completely, was one's stimulation of that system successful?

IMPLICATION 3: ACCESSIBILITY & UNDERSTANDING

And third, Chaos Theory and the more problematic Complexity have forced us to recognize a relationship between the object of study and the kind of knowledge we can claim about it. Some objects of study are not as accessible as others or at least require a different approach. Dissection works in the high school biology lab but not in the Chaos research center; knowing the parts doesn't necessarily tell us much about the functioning system. Stephen Kellert describes the problem from the perspective of Chaos Theory thus:

> "At issue here is the method of understanding: In what way does chaos theory give us understanding? By what method, by what means? And the answer is: by constructing, elaborating, and applying simple dynamical models."

Kellert maintains that there are three important characteristics of these models:

> "...the behavior of the system is not studied by reducing it to its parts; the results are not presented in the form of deductive proofs; and the systems are not treated as if instantaneous descriptions are complete." (Kellert 1993, 85)

At issue here is not how we come to know something as much as what we can possibly know about something. It's not a new problem. One can not see a molecule but one can study macroscopic chemical reactions. With the appropriate models of molecular behavior, one can infer things about molecules from the macroscopic behavior in the test tube. But one cannot prove anything about molecules *per se*. Instead, one's model is (or is not) vindicated, it does or does not seem to "fit the facts".

This constraint on the scientific enterprise was merely annoying when the epistemic limitation could be justified by limits on instrumentation: if we had the right equipment, we could see molecules, and thus we could verify, not just infer, that our models were accurate. But the study of Chaotic and complex systems gains little by looking at the components in isolation. No instrumentation will help there. Lack of verifiability is not an annoyance anymore. The "end of certainty", as Prigogine (1996) refers to it, is a way of life.

In the spirit of the philosopher Immanuel Kant, Stephen Kellert has suggested that Chaos confronts us with the "transcendental impossibility" of knowing all features of a system and therewith of fully verifying our theories or models. It is not that we haven't learned what we need to know about Chaotic systems. Instead, there is a fundamental limitation on our ability to know everything about Chaotic systems. Those limitations are built right into "our minds" like the Kantian categories of reason. Says Kellert:

> "Transcendental impossibility holds for all configurations of human abilities that retain such nonaccidental features of human inquiry as the following: being conducted by finite physical organisms, being expressed in language, and being motivated by interests and values." (Kellert 1993, 42)

Science is, of course, one such "inquiry". Scientists are included in the category of "finite physical organisms", and their work is expressed in symbolic languages (though the semantic ambiguity certainly differs between, say, French and differential equations). Anyone familiar with the work of Karl Popper, Paul Feyerabend, and other philosopher/historians of science will also recognize that "science" is a cultural artifact as well as a discipline. One could not have lectured on the Big Bang at the Vatican in the year 1200 and one cannot today do significant research without funding from sympathetic individual or corporate supporters.

STRATEGIES AND THE OBJECT(IVES) OF STUDY

David Harvey and Michael Reed, writing in the discipline of social science, are willing to make the best of the situation in Kellert's constrained universe of scientific inquiry. Taking their cue from Roy Bhaskar and P.A. Anderson, they too see the scientific enterprise as evolving along ideological as well as theoretical axes. But they turn the problem of accessibility into a hierarchy of methodologies where what we can know is a function of the object of study and of the methodology used to study it. On their view, science reveals "nested" layers of reality between which there are "broken symmetries" (Harvey 1997, 298-300). Chemistry is more than particle physics with huge particles, but the particle model has its uses. No one would deny the usefulness of neuro-psychological research in sociology, but there's more to voting habits than the 'voltages' in the brain when the voter 'throws the switch' for one party rather than another.

Given that different disciplines of science provide different but interrelated insights, and given further that the objects of study in the social sciences range from the physical artifact to the more abstract notion of cultural and evolutionary behavior, Harvey and Reed propose a "methodological pluralism" for studying "institutional life" (297). The pluralistic view is not new with Harvey and Reed, nor is it unique to the social sciences. Based on his analysis of the advancement of science (before the advent of Chaos), Feyerabend called for a "pluralistic methodology" in *Against Method* (Feyerabend 1986, 47). And Steven Horst seems to argue for a similar pluralism for the theory of mind and cognition in his *Symbols, Computation, and Intentionality* (Horst 1996). Harvey and Reed, however, present a matrix of modeling strategies that we will find particularly useful.

We don't need to delve into all the details of their matrix. A brief description of three strategies – the "predictive and statistical", the "structural", and the "historical narrative" – will be sufficient for our purposes. These three illustrate the range of models Harvey and Reed are willing to support and they illustrate how methodology must adjust to the object of study, to limitations in the object's accessibility, and to the goal of understanding. More importantly, these three strategies will resurface in later chapters when we look at how we understand software systems.

The predictive and statistical modeling strategy is best suited to the "determinant regularities of the universe", to "aggregated phenomena ... composed of additive, numerable, and interchangeable individual units" (Harvey 1997, 309). The method assumes deterministic relations between parts and wholes, between current and future events. It is intended to be predictive. As such, it is best applied to objects of study (things, events, systems, etc.) that are accessible to, and reward examination by, its methods of analysis and explanation (even if those take Chaos into account). That is, one would apply this strategy to determine the likelihood of an asteroid being displaced from its orbit and hitting Earth. One could not use it to predict the kind of orchestral music or fiction that will be popular during the week of impact.

According to Harvey and Reed, structural modeling works best for social organization, wealth distribution, etc. These objects lie between the stub-your-toe-on-it objects of mechanical engineering and the more amorphous and elusive objects of study in psychology and history, human experiences. It is organizational structure that, at least in part, makes a business run the way it does; to understand "how" a business works, one needs to understand the constraints, the parameters, in short the structure of its corporate form. Very little will be gained by reducing "management science" to biochemical laws. Much is to be gained by basing it on a scientific analysis of the complex and Chaotic social environments that are "businesses". The goal is better business management, not to understand how pervasive are biochemical laws.

As for historical narration, Harvey and Reed suggest it is appropriate when analyzing individual contributions to, for example, institutions or class struggles. These contributions have significance as experiences, not just as raw events. For those in the same cultural setting or for those who wish to 'learn from history', the goal of understanding is to penetrate into the lived experience of the individuals and to draw 'personal' lessons from those experiences. You and Christian Huygens both look like the same kind of carbon-based organism under the microscope of biological examination. Both of you see according to the same laws of optics. Only Huygens peered deeply into the Orion void and saw stars shining beyond the edge of his 'known' universe. But you can understand that experience and from it appreciate what it is like to be a spectator of the beginning of a paradigm shift.

Echoing Kellert's view of transcendental impossibility, but with a more optimistic tone, Harvey and Reed claim the structural and historical narrative strategies are best suited to "cultural products, historically specific events and collective symbolic processes" (Harvey 1997, 314). As I read Harvey and Reed, they have replaced Kellert's impossibility with a realistic pragmatism. If we can't explain voting habits through a reduction to particle physics, we can still examine that activity as a human activity using other methodologies and we can still make claims about that activity that are subject to some kind of verification.

So far, we've explored how Chaos has challenged preconceptions about scientific knowledge and how it has forced us to rethink the goals of science. Science is no longer an effort to unify "Everything" under One Grand Theory; instead, it is a cooperative effort of pluralistic methodologies pursuing sometimes ill-defined 'objects' of study. The world of baryonic particles is no more accessible than is the world of "cultural evolution" – the very small and the very large both escape our view. Chaos, too, shows that there are limits to what we can know, even when we study bounded, macroscopic, physical systems. The accessibility of the objects of scientific study isn't always direct, and even when we have access to indirect information about those objects, it isn't always clear how we are to think about them, model them, or theorize about them.

We can now understand why some proponents of the new science believe its paradigm shift extends beyond 'scientific method' and into such difficult terrain as "knowledge" or "truth". Chaos is not just a matter of finding a new set of nonlinear models for the physical sciences. Recalling the 2- and 3-tine fork example, the real paradigm shift has occurred in the realization that knowledge is a system itself, a latticework of legitimate methodologies (or "approaches", if you will) and a corresponding set of objects of study. Paradoxes like Schrödinger's cat aren't limited to quantum formalism anymore: the same object of study can be alive or dead, static or emergent, thing or experience, component or system, part or whole, at the same time.

Chaos is, of course, one 'part' of a yet larger 'whole', a 'component' of a larger 'system'. That system, Complexity, holds additional ramifications for scientific knowledge and method, and has contributed perhaps more than Chaos to the paradigm shift of the new science.

COMPLEXITY

Complexity has a number of aliases and tributaries: self-organization theory, dissipative systems theory, nonequilibrium thermodynamics, and so forth. Complexity views system behavior and evolution as the interplay of energy and structure. Such systems evolve from Chaotic nonequilibrium states into new structural arrangements, with new economies of energy production and consumption. As Harvey and Reed put it, these systems 'metabolize' energy in their environment "into increasingly more complex structuration" (Harvey 1997, 303). For Sally Goerner (also writing within the social sciences), Chaotic systems show "self-coordination and the spontaneous creation of wholes out of parts" (Goerner 1995, 25). She identifies three "messages" of self-organizing, complex, Chaotic systems:

"1. Self-organizing, self-maintaining dynamic organizations occur spontaneously far from equilibrium (they do not occur at, or near, equilibrium)....

2) Self-organization found in nonliving systems provides both a metaphor and conceptual model for living systems and supra-living systems (e.g., cities)....

3) New forms of organization emerge through the process of order through fluctuation."

Emergence is the most problematic concept in Complexity, and it is also the trumpet call that topples the walls of the classically deterministic approach to scientific knowledge. The rules of the system itself may change in a Complex system. According to Harvey and Reed dissipative systems are "boundary testing" by nature, they seek new organizational states. They are Chaotic in that they show sensitive dependence upon initial conditions and bifurcate in far-from-equilibrium conditions.

But what makes a complex system Complex is what happens when it bifurcates into a new "state". The parameters of the phase space change: "time and space coordinates, the degree of [the system's] internal complexity, the nature of the cycles governing the internal dynamics of the system, and the total environmental energy available to the system". Studying a complex system is studying the "evolution of evolution" (Harvey & Reed, 304). This drives a degree of historicity into even the most thorough and seemingly 'objective' analysis of Complex systems. We'll return to this issue in a later chapter.

I suggested earlier that at the fringes of the new science one cannot escape the age-old tensions between empiricism and idealism. By now you should certainly understand why. Objects are instantiated by structures but structures aren't things. Nor are they ideas. Systems may behave deterministically but we may not be able to understand that at the time we investigate that system. Chaos and Complexity require a revised, one might say an integrated, view of some historical schools of thought. But Chaos and Complexity are not hooligans in the classrooms of classical science, disrupting the lecture but providing no significant input themselves. Indeed, it would be unfair to leave this summary of Chaos and Complexity without sharing at least some of the optimism and enthusiasm of their adherents.

Chaotic and Complex systems force us to us to look at the world as if it were replete with different kinds of organism. The Complex world is not a jumble of lifeless, blind point masses waiting to be scattered to who knows where by who knows what. Reality is not a coincidental collection of pieces, all subject to the same laws of physics. Instead, the reality revealed by the new science is, to appropriate the image Rees uses to describe the multiverse, an 'archipelago' of evolving systems in which the 'laws' may well evolve as the system evolves (Rees 1997).

One can view this uncertainty with despair or with optimism. Goerner has clearly chosen the latter. Complexity teaches us we live in a "directed" and "opportunistic" universe, where the system and the environment in which it exists are "inseparable and coeffecting" (Goerner 1995, 35-36). This interplay affects what can be known about a system. The coarse system characteristics may be predictable; the component-level behavior may not be. The evolution of a dissipative system shows "the complimentarity of stochasticity and determinism (or, put another way, chance and necessity, novelty and confirmation)" (DeGreene 1997, 176). No matter what the price of Chaos and Complexity, they have returned to us the "mystery" of life:

> "The new understanding denies dualism, the separation of human and mind-based dimensions. We are not a mystery apart from the world but part of the mystery of the world ... the mystery is in us, of us, and more than us, all at the same time. Science (our belief in here-and-now facts) and spirituality (our sense of more-to-it-than-this) map to one physically real world." (Goerner 1995, 36)

Of course, heartwarming as that perspective may be, it doesn't tell us what we can do with Chaos and Complexity. Epiphanies are fine off company time; discussions in the cafeteria about the teleology of the Complex universe might be a nice break from talking about the latest skirmish in the browser wars. But as we shall soon see, the application of Chaos and Complexity to software development is not simply a matter of adding one more "approach" to our tool kits.

THE METHODOLOGICAL IMPLICATIONS OF THE NEW SCIENCE

The new science is a discipline. Whether working with models or with metaphors, its researchers are guided by a few common assumptions about how the world works, what can be known about the world, and what constitutes a 'true' or at least 'reasonable' proposition about the world's workings. The 'world' may no longer be partitioned into components whose behavior simply adds up to the behavior of systems. Not all aspects of the world may be knowable, and not all may be knowable in the same way. Statements about Chaotic systems must be qualified with some rider clause such as "so far" because Chaos drives Complex systems, and as Complex systems evolve, so may their structural rules. The new science is certainly no panacea for uncertainty and impatience.

Nor is it without risk. Models are inherently imprecise. All modeling is selective - it highlights some aspects of the available data at the expense of other aspects (Gleick 1987, Morrison 1991). Models, especially Chaos models, also may be inaccurately applied: while the available models have been well documented (e.g., Williams 1997), so have the difficulties of their application. And since Chaotic systems don't exhibit predictable, linearly evolving behavior, selection of data points is crucial to the success of the model.

Harvey and Reed address a related issue when they contrast models with theories:

> "Models, as opposed to theories, are well-formed metaphors and analogies. They do not claim to express the truth of the world, but merely to provide heuristic insights. While theories claim to actually explain reality, models are only partial, fictitious constructions." (Harvey 1997, 309)

One might argue with their choice of "fictitious" to describe models, but the idea that models are metaphorical and heuristic should cause us some concern. It seems to me that this is where the new science is perhaps its own worst enemy. It supplies us with emotionally charged metaphors. It speaks of far from equilibrium states, of turbulence, of strange attractors, bifurcations, and emergent entities. These are easily applied to situations that are not Chaotic or Complex in any legitimate sense – a turbulent family relationship is not necessarily a Chaotic one. Several authors have suggested that the metaphoric applications are actually the more prevalent use of Chaos in the social sciences (e.g., Ruelle 1991 and Elliott 1997). Such use is not without risk.

Even in legitimate applications, a metaphor may be misleading. There is significant evidence that brain activity and thought itself have Chaotic characteristics. There may in fact be strange attractors in brain activity. But the notion of an attractor leads us to envision a locus, say a particular neural net, when the attractor may well be a higher order, systemic entity that does not guide activity, it instead creates it.[14] A slight nuance of a heuristic metaphor can mean the difference between a decrepit model and an adequate one. And neither models nor metaphors intrinsically take into consideration the historical limitations of studying Complex systems, where what one understands of the system may be an artifact of one's temporal, social, or cultural "location".

THE MORAL CHALLENGE FROM THE NEW SCIENCE

The problem is, the software industry simply can't ignore the new science and hope it goes away. Chaos lurks in our brain's neural patterns, in our heart and breath rhythms. It appears in our voting patterns, in the ecosystem that maintains our existence as voting organisms, and perhaps even in the universe we inhabit as organisms. Complexity is almost an ideology in the way it forces us to radically reexamine how we think about the world and how the world 'works'. The new science forces us to take a critical look at our methods and their rationales, if only to determine to what extent systems might be Chaotic or Complex. And if there are systems that may be Chaotic or Complex, it provides us a way to analyze them and, maybe, to help manage them.

Everything we've learned about Chaotic and Complex systems indicates that they may be impenetrable to the quality profession's deterministic tool set of reliability analysis, root cause analysis, and piece-part quality assurance. Where software quality assurance expects linear changes in state, Chaos and Complexity instead reveal oscillations, bifurcations, even 'quantum leaps' to new states. Where the industry deterministically expects future behavior to be based on current known behavior, Chaos and Complexity drive a wedge between whatever pieces we can describe and the dynamically changing structures that comprise them. Quality assurance for a product that was potentially a Chaotic or Complex system would be closer to roulette than to regimen. The product might pass inspection and test on Monday, succumb to some force that pushed it far from equilibrium on Wednesday, and by Friday have bifurcated into a new system, with new behavior characteristics that had not been inspected and tested. It would make setting up the latest Beta of the NT™ operating system look like child's play.

"Continuous improvement", "process capability", and other such tagwords that reflect the software industry's long-term commitments to "quality" and "return on investment" and "efficiency" would certainly have new meaning. They might have no meaning at all. How could we know that a specific technique used in the development process had been effective when what was produced was dynamic and potentially unstable? How could we define "effective" in the first place, given that the future behavior of a Chaotic system isn't deterministically predictable from the initial conditions of the system?

To return to a point made in the previous chapter, today we cannot prove that a given approach is "better" than another because we lack the criteria for proof. "This [enter a technique here] works for us" may be an accurate description of how we feel, but it is not a verifiable statement about the technique: we simply don't know whether that technique or some other factor caused the success of our development effort. And we don't have that information because we often overlook the "system" of development. We don't consider all the factors that collectively contribute to the production of the final product. We don't consider all the interrelationships between factors. And we don't consider the various characteristics of the "systems" that are represented by the different development and business and product environments. We can't justify our methods because we haven't completely analyzed the entities to which we apply the methods.

Suppose for a moment that future research shows there is a class of software product that is in some sense a Chaotic or Complex system. Would we approach these products any differently? Would we even recognize them as different? After all, we don't apply methodologically legitimate approaches today; technique X might "work" just as well in Chaotic systems as in non-Chaotic systems, or fail, we wouldn't know which. What's the difference between not knowing whether your approach is effective, accurate, and precise, and applying that approach to an entity where effectiveness, accuracy, and precision may no longer have any meaning? For all intents and purposes, nothing. And that should frighten anyone who calls themselves a software professional, or who ships products with accompanying warranties, or who participates in software quality assurance, or who uses words like reliable, correct, good, etc. to describe software.

The new science invites – one might say morally obligates – software professionals to apply greater regimen to our current behavior in order to determine if we need worry about Chaotic or Complex systems in software. Such systems still need to be accurately described. They still need to be measured precisely. Even a Chaotic model is, or is not, effective at describing the structure of the system. So if there are such systems to be found amongst the software that increasingly controls our lives, we need to hone our methodological skills enough to understand them to whatever extent possible. The new science, despite its "mysteries", is still a discipline; it is not an excuse.

CHAOS, COMPLEXITY, AND SOFTWARE – A LOOK BACK AND A LOOK AHEAD

We should have recognized software development in Kellert's list of conditions for Transcendental Impossibility: developed by humans, couched in language, influenced by intentions and values. We should by now suspect that the act of production and the entity produced – the development effort and the software product – may be more than a linear and predictable production effort ending in a stable "thing" independent of the developers and users. Software development and software products are cultural artifacts; they are based on human needs, human design and intentions, human skills. They are culturally, socially, and historically conditioned.

But are they Chaotic and/or Complex? I will try to answer that for some classes of software in later chapters. Before we embark on that analysis, let's collate the results thus far. Although I introduced effectiveness, accuracy, and precision in the context of process capability in manufacturing, here I will describe them from the classical scientific perspective. This will highlight the difficulties software professionals face, caught as they are between the classical and new sciences.[15]

EFFECTIVENESS:

- Classical science: A scientific theory or model is effective if it succeeds in predicting, or succeeds in elucidating for a class of objects, the occurrence, structure, and/or components of a class of objects or events.
- Software: We do not today meet the classical scientific criterion of effectiveness for any of the following reasons:
    - We do not have incontrovertible empirical evidence that an approach does what it claims to do
    - We do not or cannot deduce its effectiveness from scientific or philosophical principles
    - In formulating and defending approaches we do not address all the intricacies of the development effort and/or the product.
- Chaos: Chaotic system must be examined as systems, not as aggregates of component parts, so the effectiveness of a model or theory depends upon its ability to:
    - Elucidate the relationships between components not the components themselves, and/or
    - Identify and (loosely speaking) predict structural behavioral patterns.

ACCURACY:

- Classical science: Due to assumptions about reality and/or due to perspectives on the goals of science, the objects of study are assumed to be stable, non-emergent entities that follow (in whole and in part)

linearly predictable patterns of behavior through time; they have historical trajectories as it were.
- Software: Unable to define and determine effectiveness, the software industry is correspondingly unable to prove that a given approach is accurate for a specific set of conditions.

    - The situation is exacerbated by the fact that in many cases each development effort is a "new" effort, so in cross-project comparisons the validity of the comparison is questionable or requires more detailed analysis of all contributing factors than is generally undertaken.

- Chaos and Complexity: The new sciences, on some interpretations, invite a methodological pluralism that provides a hierarchy of methodologies; analysis adjusts to the object type and to the goal of understanding the object. Accuracy is less a matter of reproducible results than it is of completeness and appropriateness – a variation on the requirements of coherence with other theories ("connectedness") and applicability to other similar cases ("extendability") in current scientific theoretical evaluations (Coles 1997, 11).

PRECISION:

- Classical science: The results obtained with a theory or model must be the same across applications, within some accepted tolerance.
- Software: A given approach is precise if it can produce similar results across multiple applications. This requires a detailed knowledge of the context of application and the factors involved in "success", as well as a definition of "similar" based on empirical measurements or rational (categorical, philosophical, or scientific) classification.
- Chaos and Complexity: In Chaotic or Complex systems, it may be impossible to specify exhaustively and precisely all the relevant state descriptors, and even if this can be achieved future state descriptor values may not be predictable. Precision is at best approximation, and

in Complex systems it further suffers from historical limitations because Complex systems evolve.

Looking ahead, this examination of Chaos and Complexity has also identified some guidelines for investigating software systems.

THE PART-WHOLE RELATIONSHIP

- The new science: There are "broken symmetries" between the 'layers of reality' revealed in a methodological pluralism; parts may not behave according to the same laws as wholes, components may differ behaviorally from systems.
- Further investigations: Identify the 'parts' of a software system, especially those that may contribute to or exhibit Chaotic behavior. It is to be expected that such a "system" will contain "parts" that are otherwise treated as independent entities, such as "requirements", "code", and "product quality".

FAR FROM EQUILIBRIUM DYNAMICS

- The new science: Chaotic systems appear "normal" at equilibrium but radically change their behavior, and perhaps even the structural relationships between their components, when perturbed into a far from equilibrium state.
- Further investigations: Define and distinguish for software equilibrium and far-from-equilibrium states, determine the mechanisms of disturbance, and the behavioral and/or structural changes that may occur.

EMERGENCE

- The new science: A Complex system may pass through a Chaotic phase and change its structural identity. Analysis of such evolving systems is possibly constrained by the spatio-

temporal/social/cultural/historical location of the observer. There is an historicity to understanding Complex systems in general, and where language or values are involved we the observers view such systems from within a social or cultural horizon.
- Further investigations: The roles of language and values must be explored to determine their impact on the characteristics of the software "system" and whether there are, in some sense, "emergent" features of such systems. The relationship between the social or cultural horizon and our understanding of a software "system" also needs to be explored.

CHAPTER 4:
CHAOS IN SOFTWARE?

> ...round he throws his baleful eyes
> That witness'd huge affliction and dismay
> Mixt with obdurate pride and stedfast hate:
> At once as far as Angels kenn he views
> The dismal Situation waste and wild,
> A Dungeon horrible, on all sides round
> As one great Furnace flam'd, yet from those flames
> No light, but rather darkness visible
> Serv'd only to discover sights of woe,
> Regions of sorrow, doleful shades, where peace
> And rest can never dwell, hope never comes
> That comes to all; but torture without end....
> John Milton, *Paradise Lost*, I, 56-67

Such is the view that greets Satan and his fallen legions on the plains of Hell after their rout from heaven. Sound familiar? Does it bring to mind your last custom software development project three weeks before ship date? How about that Web-enabled publishing effort your IT group took on for the Business Unit That Eats Programmer-Analysts For Lunch? And we won't even mention maintenance of complex distributed software systems in technologies where the software is aging almost as quickly as the maintainers are.... Is there solace to be found for those of us who don't hold stock in companies selling ulcer medication and anti-depressant drugs?

Some have suggested that in order to avoid chaos in software development we must first embrace Chaos. Ignoring the claims of people like Capers Jones (who argues that software development should be treated just like manufacturing) authors such as Olson or Bardyn and Fitzgerald maintain that software development and software project management exemplify Chaos.

They insist we must accept Chaos as a way of life. They even suggest some steps we can take to manage Chaos in projects and in products, today, right here in Cubicle City.

JUSTIFYING CLAIMS OF CHAOS IN SOFTWARE

The material presented in the previous chapter should leave us wary about such claims. Chaos (to say nothing of Complexity until a later chapter) should not be underestimated. Managing Chaos, if that's possible at all, requires a thorough understanding of the components of the system, the trajectories over time of components within the system, and coeffecting influences. Just what does it mean to say that "software" is Chaotic? How would we justify that claim, let alone account for it in our next project plan?

The issue of justification is critical. As I've pointed out before, the new science provides powerful metaphors, not the least of which is "Chaos" itself. Using a metaphor or analogy only requires poetic license; applying a mathematically sound Chaos model to data requires a more substantial understanding of the system from which the data was obtained. Metaphors and analogies usually highlight a few qualities of the two entities being compared, while the ideal scientific or philosophical analysis accounts for the qualities that the metaphor ignores. This is not to disparage the heuristic use of rhetorical devices. Nor am I claiming that because some position has the trappings of a scientific or philosophical analysis it therefore is. One can apply a model indiscriminately, making the data "fit" the model rather than the other way around. And it is an occupational hazard of philosophers sometimes to ignore reality when there is speculating to be done.

I am, however, cautioning us to 'look at the big picture' when someone claims software development or project management is Chaotic. For example, Sivak suggests that software test engineers can "use" Chaos Theory in testing by providing random inputs to the software program under test (Sivak 1998). His notion of 'using Chaos' rests on a supposed analogy between the "sensitive dependence on initial conditions" in Chaos and the inputs to the software program in testing. Sivak assumes, he does not demonstrate, that software programs can be Chaotic.[16] He also assumes that these inputs act like initial conditions in a physical Chaotic system. But in what sense do the inputs have a nonlinear evolution in the system? The answer to that question is crucial to the claim that random inputs incite software programs to exhibit Chaotic behavior; the answer is no where to be found in Sivak's discussion.

Before we can legitimately call something Chaotic, we need answers to the following questions:

4.1 Are the key characteristics of Chaos present?

4.2 Are we examining a system or a collection? A system comprises interdependent components and pervasive influences; there are laws or patterns of behavior that affect and are affected by the components. A collection, on the other hand, is merely a coincidental and unrelated group of pieces, corralled by a term or concept but not by any actual affinity or relationship.

4.3 Can the system be modeled mathematically as a Chaotic system?

4.4 If the system components are such that they elude current empirical analysis (e.g., thoughts), can the claim of Chaotic behavior be substantiated formally by applying evidence or hypotheses from other relevant disciplines?[17]

4.5 Even if we can describe this system only through metaphors and analogies today, is it conceivable that at some point we can substantiate our claim of Chaotic behavior at a later date (perhaps after research in other fields has been completed)?

The first and second questions above are methodological requirements for analyses of Chaos. If we don't apply all relevant aspects of Chaos Theory, and if we don't apply tenets of Chaos to systems, then we've merely appropriated some jargon from the new science and created yet another approach. The third and fourth questions above explore options for validating the claims. The fifth question accommodates the fact that the new science is new and those who apply it beyond the physical sciences are doubly challenged to break new ground with a tool that has not yet proven its temper on that particular soil.

My chapter title ends in a question mark precisely because we have two options for justifying claims about Chaos and because we may have to wait for further developments in other disciplines. We'll soon see that there are many features of software development and software development projects that could contribute to Chaotic behavior. And we'll discover that to understand software "systems" fully we'll need to draw from research in other fields.

Much of that research has yet to be performed. I should point out that even once completed the results of that research and the conclusions the software industry draws from it might not be universally convincing. David Ruelle has suggested that we should not bestow the title "Chaotic" on any system until we can specify mathematically the dynamics of that system. In Ruelle's view, "Chaos" is little more than an intriguing metaphor in the case of complicated but not well specified systems: in finance, economics, the social sciences, and in much of biology talk of "Chaotic" systems is either premature or simply wrong (Ruelle 1997). The software industry should be prepared for the possibility that we, like the social sciences from which we draw data, will be accused of "scientific philosophy" rather than science by those who take Ruelle's position (Ruelle 1991).

As we saw in the previous chapter, that threat has certainly not dampened the enthusiasm in social science for analyses that use the structural components of Chaotic systems rather than the mathematical models of such systems. I believe the pluralistic methodology of Harvey and Reed implicitly rejects Ruelle's limitations on the "use" of Chaos, and I further believe that Ruelle is making explanation in the hard sciences the exemplar of "adequate" knowledge. Ruelle's position merely admonishes us to proceed cautiously when we analyze software systems. In 1987, James Gleick showed us that Chaos is ubiquitous, but he never strayed far from physical or applied science. In his pioneering book *Exploiting Chaos* (Olson 1993), Dave Olson took Gleick at his word and proclaimed that many aspects of software development are Chaotic. While I obviously agree with Olson's proclamation, I believe he fails to justify it on either option 4.3 or option 4.4 cited above. It is that justification (I did not say "proof") we will pursue in this and in the next chapter.

DIFFICULTIES MOVING FROM PHYSICAL CHAOS TO CHAOS IN SOFTWARE

Analyzing software, software development, and software development projects is rather like trying to reconstruct the history of Stonehenge. There are some obvious artifacts from one period or another, but some features are overlaid on other features. While one can trace the development of the monument from a simple earthen ring structure to the more complicated structures we see today, Stonehenge does not represent a radial 'pagan sprawl' from some central point. In fact, some of the inner structures postdate outer structures (Aveni 1997, 63-4). The precise astronomical purpose for the various structures is still debated, but it is obvious that later architectural additions had to work with the horizon circumscribing the hill and with the field of view provided by (or, conversely, limited by) previous structures already standing. Looked at over time, the evolution of the monument occurred as interrelated efforts, separated by time and culture, but no addition was uniquely conceived and executed. The extant structure affected the new construction; the pre-existing whole affected any new part.

Similarly, software development doesn't occur in a temporal or a cultural vacuum. What came before affects what occurs today – developers learn from their mistakes, last year's earnings influence this year's project budgets which in turn affect what can reasonably be done in the project (as opposed, alas, to what is often listed in the requirements). There are, as it were, concentric rings of influence on the product, the developers, and the project. Each successive ring describes a broader, more abstract influence, from the 'immediately personal' experience of the developer, to the social interaction between the development team and the customers at the conference table, to the larger corporate vision that identified the visitors as customers, and so on. And of course the rings, these spheres of influence, are not independent of one another. For example:

The code is created by the developers, each with their own individual habits of coding and with their own (hopefully similar) understanding of the product requirements. But this sphere of influence is affected by a larger sphere characterized by the constraints of the budget, schedule, and technology.

The developer's understanding of the requirements is affected by the physical and social structure of the organization – the entire team sits together or does not, they communicate regularly with product managers and sales people or they do not.

The social structure of the organization is affected by the "culture" of the business. The internal IT customer is across the hall and attends all meetings; on the other hand, in systems houses the customer is a silent, faceless, unknown buyer who probably doesn't much care about the product except as a platform for some application they do care about.

In turn, the business "culture" is affected by sociological pressures in the society at large. Market pressures differ from one country to another (see, for example, Hammond 1996). Exactitude not timeliness may be important in one country whereas in another first to market is more highly valued.

What is true of the development project is also true of the individuals involved in it. Their actions and decisions and reactions to their environment are driven by several spheres of influence. Their experiences, their personal goals, their understanding of the product and the business combine to circumscribe their own personal 'horizons'. One developer is a recluse, concerned only with working on new technologies and with completing the assigned tasks. Another developer intends to be a development manager, and the project interpersonals are as important to this person as are the product interfaces. In contrast, the project manager lives in a world of costs and time and resources, not code and tools and tests. This particular project manager is constantly under pressure to increase the scope to please the clients and at the same time cut development costs. At least the project manager can take solace in the fact that this impossible situation ends when the last line item on the project plan is complete. The VP responsible for the project may be sympathetic with the project manager's plight, but the VP's horizon is broader. It includes multiple projects, the short term and long term corporate business goals, and an appreciation for impacts across projects.

So to our list of influences on software development, we should add another: individual personal perspectives or horizons. The overlap between spheres of influence and the interplay between individual horizons are just one reason why taking physical Chaos Theory and applying it to software development is problematic. It is difficult to specify a system that involves vital components and social influences because it clearly is not bounded the way a material system is. (This is, of course, one reason why Ruelle cautions against premature use of "Chaos" outside of the hard sciences.) Physics studies events; software development is a combination of decisions and events. Physicists study river rapids; development and project managers have to navigate through them. Are students of Chaos in software faced with the daunting task of studying a "system" comprising the entire river and the raft and the rafters? Transcendental impossibility indeed!

To illustrate, let's look at a relatively "simple" system. In a discussion between two or more people, what is a linguistic influence and what is a psychological or sociological one? Words don't have meanings, they are meaningful to speakers and writers and readers. A discussion often evolves indeterminately based on the interplay of what is spoken, what is heard, and how each participant responds cognitively and emotionally to the discourse (see Figure 4.1). If one wanted to analyze a discussion as a potentially Chaotic system, one would have to account for, and determine the relevance of, the components or characteristics of the discussion: the semantic, syntactic, emotional, imaginative, and cognitive components, each participant's knowledge of the topic, body language, etc. What appears at first to be a simple conversation turns out to be potentially very complex and difficult to analyze or to model.

If one could snapshot a conversation and map those states to sociological, psychological, and linguistic influences, one would see differences between what is said and why it is said. The "course" of the conversation (the trajectory of the semantics) occupies different zones than does a participant's cognitive and emotional state. Each affects the other as the conversation emerges between speakers. Ignoring intentions and treating such a system as a set of audible events is possible, of course, but it is also clearly an oversmplification of the system.

The difficulties in analyzing non-physical systems do not stop with the 'complexity' of influences on such systems. What are the components of such systems and how are we to understand their initial states? The start of a development project is not like breaking a rack of billiard balls (though it might feel like that sometimes). The initial conditions and the "trajectories" of the components are radically different – they are qualitative, abstract, and even sentient in the one case, lifeless, material, and quantitative in the other. For billiard balls "equilibrium" is easy to define as a state of rest; but as any engineer or project manager will tell you, there's no such thing as a state of rest in software development. And if equilibrium is hard to define for software development, how is "oscillation" or "turbulence" to be defined? How can one claim that a system is Chaotic when the attributes of Chaotic behavior remain undefined and the structural components of the system aren't accessible to "normal" empirical analysis?

Figure 4.1 - Mapping a conversation.

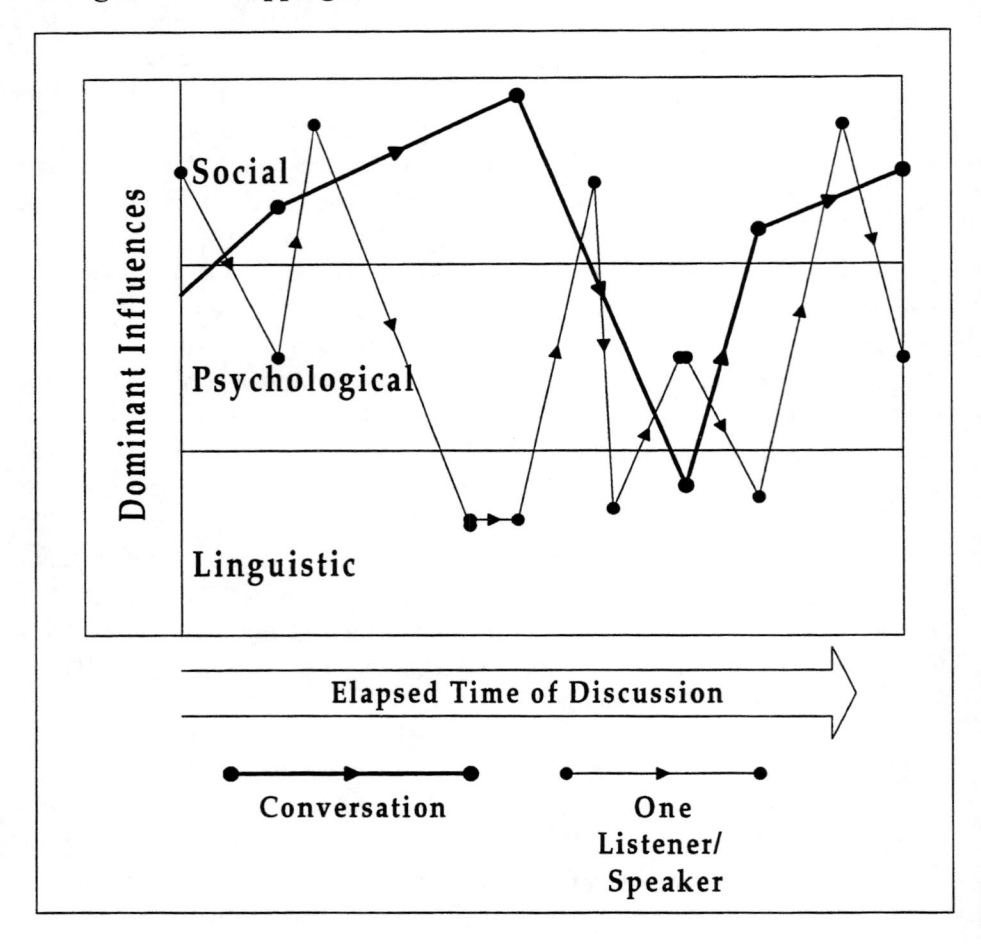

Dominant Influences: Social, Psychological, Linguistic

Elapsed Time of Discussion

Conversation ● ──▶── ●   One Listener/Speaker ● ──▶── ●

One might decide to eliminate aspects of the system to make it more manageable. One could, for example, ignore what the project team are thinking or what they believe and merely look at what is said and done. In effect, we'd be treating the team members the same way test engineers often treat software – as a "black box". This reductionist approach might help identify project-wide trends, so it could prove useful in analyzing a software "system" defined as the course of the development project. It might be an appropriate perspective for examining developer productivity (measured as the event of 'lines of code or function points completed'). It might prove useful for determining whether there is some potentially Chaotic relationship between accrued cost, elapsed time, and product completion.

But the reductionist approach also has some severe limitations: it clearly cannot penetrate into the experience of developing requirements, of writing code, of maintaining a product. To rephrase a point made earlier, while we "know" that sentient hominids can be studied as if they were giant billiard balls, we also know that "we" are more than our motions and molecular structures. We refer to software projects as development "efforts" not "events" because we are all too aware of their cognitive and emotional aspects. So what is to be done here? It appears we cannot identify and comprehend all the interactions between all the spheres of influence in all cases, yet we don't want to arbitrarily circumscribe our field of view just to simplify the investigation.[18]

I suggest that we ply the same course Harvey and Reed took when faced with a similar dilemma in the social sciences. As we have seen, there too the researcher can view agents merely as bipeds in motion or the researcher can attempt to probe into their thoughts and feelings and values. Migrations can be viewed as (possibly deterministic) physical motion. To understand voting habits requires an analysis of social structures and the psychological behavior of 'classes' of individuals. These are also required if the goal is to understand history, but they often must be supplemented by that personal, interpretive kind of understanding that allows us to say things like "I understand why you did what you did." The sociologist adjusts the methodology to suit the "object" of study, where "object" refers both to the thing studied and to the goal of the investigation.[19]

When we search for possible systems in software, we find that our "objects of study" are fewer than those that beckon social scientists but they do exhibit a similar structural relationship. The sociologist is confronted with multiple "overlapping" or "nested" systems, from the family unit to the family of nations, where the larger systems include the smaller. Whether we look at "software" as a tool or as a task, as a product to be used or as a project to be completed, we find that we can identify systems that function as parts of larger systems. Let us first look at some potential systems in software, identifying some of their components and some of the influences and conditions that affect them. Then, later on, we'll change our perspective from the variety of objects (things) of study to the variety of objects (goals) of investigation and the strategies we can use.

NESTED SYSTEMS

From the perspective of software-as-product, the executable code and its execution on a processor is the most primary (or fundamental, or central) system. More inclusive than the primary system is the execution of the software 'package' by a particular user for a particular purpose. The context in which this second system occurs is larger: to the physical execution environment of the code-and-processor it adds the goals and values of the user. The context provides further differentia between systems at this level. The goals and values of a test engineer during build integration testing differ from those of the end user during acceptance test. And an older product may be executed in a different environment from a new product: changes in user expectations and requirements, changes in platforms, and changes in the work flow in which the software package is used may cause the product's perceived performance to change.

Looking at "software" as the production of software rather than its execution, we find a similar set of nested systems. Central to other systems are the very constrained systems of developer activities, such as reading the requirements and imagining what needs to be done. Moving outward from these 'individual' systems, one might consider requirements, design, and code reviews as miniature 'social' systems that are themselves comprised by the larger social system of the development project as a whole. The development project itself can be examined as a bounded system, influenced by organizational structure and corporate priorities and constrained by time, resources, and budgets.

Let's examine several of these systems in enough detail to identify their major components and where the spheres of influence come into play. Then we'll look at the strategies available for analyzing them. We'll approach systems from both perspectives – software as product and software as production effort – and within each category we'll briefly describe several systems.

SOFTWARE AS PRODUCT
The Code Executed in a Hardware Environment

This system is bounded by linguistic, physical, and structural constraints. The linguistic and structural components come into play here because the executable is not a typical physical entity: the bits in memory are simply the material instantiation of the primarily linguistic and structured entity that is the "code". The content of the program remains implicit or latent until it is executed on the physical hardware, and the hardware environment responds to the content of the program by executing its "statements", acting in accordance with the structure of the program. Yet it also constrains the effects of the statements and it can affect the structural flow of the program through resource management or error handling. This system has, minimally, the following components:

- Linguistic components:

  - Directives
  - Variables and their declarations
  - Calculations[20]
  - Inputs and outputs
- Structural components:
  - Code structure reflecting operations intrinsic to the program (the code 'paths')
  - Code structure reflecting operations extrinsic to the program (the product 'functions')
  - Error handling (timing, security, data integrity, resource request failures, etc.)
- Execution environment components:
  - Operating system
    - Memory management policies
    - I/O processing policies
    - Interrupt management policies
    - Resource management policies, including scheduling, locking, and SMP designs
  - Other programs competing for processor resources

- Processor speed
- Peripherals
  - Bus bandwidths, protocols, etc.
  - Response speed, capacity, etc.

The Product Used for a Purpose by a User

To the set of influences and constraints operative at the level of the executing code, this system adds psychological and sociological factors. The "context" in *this* system is more than what is saved in some data structure or on a stack. It now includes the user's values and goals which in turn reflect the user's prior experiences with software products, the 'work flow' in which and for which the software product is used, and the workplace and business environments. A harried office environment can increase a user's impatience waiting for an output: in the execution environment the product is behaving to specification but in the use environment it is sluggish. On the other hand, users jaded by a long history of decrepit applications on their desktops may be more inclined to overlook product functional deficiencies than those new to the workplace or those who sit in the QA group.

The 'personal' context is an essential part of the system in this case. Not only does it affect how the experience of using the product is perceived; it also provides quantitative and/or qualitative criteria for describing the system. We can illustrate this with a comparison (admittedly simplified) between integration testing and acceptance testing.

The test engineer compares product functions against all requirements. These may be quantitative or qualitative:

- Functions are or are not performed to specification
- Reliability under normal operation is/is not to specification
- The program is/is not fault tolerant
- Performance under normal and excessive loads is/is not to specification
- Displays, outputs, and calculations meet Correctness criteria

The acceptance tester compares product functions against workflow-related requirements and personal expectations. These are more often qualitative than quantitative:

- The product "fits" the work flow
- It meets the user's ease of use expectations
- The perceived performance of product is satisfactory
- Response time is satisfactory
- Outputs, graphs, calculations, etc. are correct

An Older Product Used in a New Environment

In addition to the components and influences operative in the previous two systems, this system is affected by time. The product has endured in the marketplace or workplace long enough that the execution environment, the use environment, and the business environment have changed substantially. Many development efforts take so long to complete that this description could be reasonably applied to 'new' products, too. But for our purposes we can focus on tried and true products that, over time, become tired and trying.

New personal contexts will generate new expectations about product features and performance. New business environments make new demands of workers and their software tools. New execution environments produce new resource and memory management policies, introduce new timing issues in the form of higher bandwidths and new protocols, etc. The aging product may no longer be fault tolerant; it may have low reliability or availability. It may stumble when applied to new problems and new tasks in the workplace. On certain platforms it may fail completely. The users are not impressed: after all, the software product is just a "tool", so why doesn't it work as well here as there, as well now as back then?

SOFTWARE AS PRODUCTION EFFORT
The Act of Coding

It is 3 PM in the afternoon on the Friday before the first drop is due in the quality assurance lab. The cubicles and quads and 'working lounges' are filled with sounds of keyboards and the hum of people chattering about implementation issues. An astute listener hears more than the noise. The tempo of typing in this cube reflects the pressure this developer feels. Across the room there is a crescendo of keyboard clacks, followed by a ceremoniously hard slam on the Enter key, signaling the demise of showstopper bug #674. From the coffee brewers drips the hallowed Elixir of Endurance – for the tenth time since noon. The brewers were due for cleaning this week, but the development manager wasn't born yesterday, and they were descaled two weeks ago.

The expectations of some product and project managers notwithstanding, code development is not a kind of autonomic response that occurs in the presence of a project plan: the act of coding is a system affected by linguistic, personal, psychological, and sociological factors. Looking at these in reverse order, we can identify some (though by no means all) typical and important components and conditions 'by sphere'.

- Sociological Conditions:
    - Time constraints on the project
    - Accuracy of direct labor estimates
    - Accuracy of duration estimates
    - Accuracy of dependency calculations
- Budget constraints on the project
    - Availability of tools, hardware, training
- Project (Department, Corporate) Priorities
    - Good or Fast or Cheap when you can have only one
- Social environment
    - Support from management
        - Developers are insulated from anything that might lessen their concentration
        - Coffee brewers were cleaned
    - Social / Structural interaction
        - Developers work alone
        - Peer reviews or desk checking
        - Frequent meetings with users
        - Specifications are available
            - Developers are fluent in the language used to write the specifications
- Psychological Conditions:
    - Worker contentment
        - Salary

- Working conditions (lighting, noise, ambient air temperature and purity)
- Worker Focus
  - No administrative interrupts
  - Flex time, etc.
  - Family life stable
- Response to pressure, motivators, etc.
- Neuro-biological conditions
  - Coffee
  - Individual biorhythms, blood chemistry
  - Sleep/fatigue
- Personal and Linguistic Conditions:
  - The developer's experience with the platform, product type, end users
  - Knowledge of the programming language, ability to create logical/functional structures
  - Assumptions about user needs (work flow, expectations) based on reading specifications and talking with users
  - Ability to project (imagine) affects of his/her code on other aspects of the product she/he is not writing
  - Ability to project (imagine) affects on the code being written from processing environment, demented testers, users

The Development Project

Even though the previous lists of conditions and influences were not exhaustive, they did indicate how complicated those systems were. One would therefore expect the system that is the development project to be even more complicated, since it contains the systems mentioned previously as well as others. Fortunately, we don't need to plunge into a full analysis of the development project. For our purposes it will be enough to point out some of the structural characteristics of the development project as "system":

The project is a mini-society experiencing a limited form of evolution. It has a linear history reflected in the project plan. It has social hierarchies of managers, leads, and 'grunts'.

It experiences cultural change, as the 'project resources' come together, work together, and become a 'team', and as the requirements are refined and the product takes shape in the source code.

The project forces a social structure of its own on the people involved and on the production process: there are phases, reviews, and milestones.

The project combines other systems into a superstructure of tasks and deliverables and to those systems it applies a set of evaluative criteria (usually time and money).

By assigning time and budget to tasks the project creates a context in which the other systems must operate.

The developers have the training and the experience and the hardware they need, or they don't. They have enough time to really think through design problems or they have only enough time to code.

Considered structurally, the development project system is an active force in its own growth. It is subject to external influences, to be sure, but it also affects which of its constituent system's influences can be effective in their original contexts.

STRATEGIES FOR ANALYSIS

The previous section described some systems that the software industry might examine for traces of Chaotic behavior. The catalogue of systems is by no means exhaustive, and by itself it does no more than alert us to the myriad of influences, conditions, and interrelationships that we would have to consider in a detailed analysis of any particular system (some of which are illustrated in Figure 4.2). Based on what is already known about Chaos in social systems and in the psychology of the individual, we can be certain that many of these components could in certain combinations and under certain circumstances contribute to Chaotic behavior. And for that reason, the previous lists, incomplete as they may be, outline the start of a major research effort the software industry must undertake if it wants to definitively understand where, when, and why Chaotic behavior can occur in software development and in the use of software products. To reach that goal will require much Chaos research, most of it yet to be initiated, in sociology, psychology, and mind.

Systems and Spheres Matrix	Sociological	Psychological	Personal	Linguistic
Code executed in the CPU				■
Code executed by a user for a purpose	▓	■		▓
Developer writing the code	▓	■		
A requirements review meeting	■			

Figure 4.2 - Spheres of influence as they impact sample "systems".
This matrix illustrates how the spheres of influence affect different software systems. Darker shading indicates greater influence.

It will therefore be some time before we can remove the question mark from the title of this chapter. In the meantime, another methodological challenge awaits us. We still need to understand how to analyze such systems, and what the different methods of analysis available to us can reveal and what they might conceal. For that understanding, we can turn once again to the position taken by Harvey and Reed and discussed in the previous chapter.

We share with sociologists a number of ways to analyze these systems. We can look at events, structures, and/or decisions; we can treat the object of study as thing, as a set of relationships, and/or as the voluntary expression of individual experiences and volitions. We can collect incidence data irrespective of the who and the why. We can examine patterns of behavior, looking at the interdependencies between incidences of some set of events but not necessarily probing the inner experiences of the agents involved in these events. And we can probe the experiences of the agents themselves. These three perspectives or strategies roughly correspond to the three "modeling strategies" from Harvey and Reed discussed in the previous chapter. Figure 4.3 summarizes the relationships between the strategies and the spheres of influence discussed earlier. We'll briefly look at each strategy in turn.

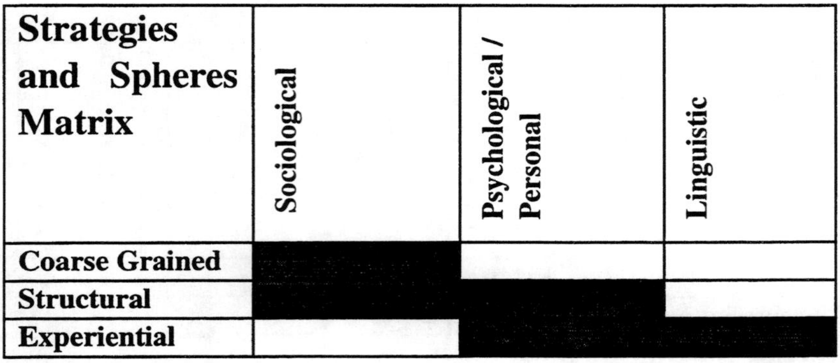

Strategies and Spheres Matrix	Sociological	Psychological / Personal	Linguistic
Coarse Grained	■		
Structural	■	■	■
Experiential		■	■

Figure 4.3 - Mapping strategies to spheres of influence.
Not all strategies are effective for probing all spheres of influence. Shaded areas in the matrix represent the best fits.

THE COARSE GRAINED STRATEGY

On this model, we look only at physical events, at changes of states, at "accidents" if you will. An event can be the fact that there is a disagreement over the meaning of a requirement. It can be a budget overrun. It can be the completion of a document or the completion of a project task such as coding a function. It can be the classification of some behavior of the software as a "defect". In all cases, the knowledge, desires, felt constraints, etc. of the agents involved in the events are generally ignored. Since the 'experiential context' of the event cannot be known in the same way an event can be known, this model makes only limited use, if any, of that context. On this model, the world is a "coarse grained" collection of things in motion, a chess board in which states are as simple and as well delimited as black and white squares and no one looks for finger prints on any playing piece.

For example, we can view the developer's production of code as a black box, measuring the development time used per function point produced. We might consider the type of product being encoded as part of the state changes we are examining. But we won't consider the developer's IQ, the university the developer graduated from, or the developer's daily metabolic cycles and hourly blood chemistry. After sufficient amounts of data have been collected we can make the kinds of predictions that fill the pages of Software Productivity Research articles (Jones 1997). Except for its weaker predictive value, the coarse grained strategy is similar to Harvey and Reed's "predictive" or "statistical" strategy.[21]

The coarse grained strategy is best suited for the common-sensical material realm and for coarse behavior characteristics when the peculiarities of the agent(s) are less significant than the action(s). This model can be expected to generate useful data when it is applied to such events as project variances computed as more than 5% of existing budget, or number of defects per development phase (where for convenience we assume that "defect" has an objective definition). It can also be applied to classes of behavior in large contexts such as cultural or social behavior, as when comparisons of development project data are made across projects and across industry segments.

THE STRUCTURAL STRATEGY

We can also look at an event (say, the production of code) from the project perspective, seeing it as an extension of the methods used to generate the requirements and the design. On this view, the developer is more than a manufacturing biped; he or she is an active agent in a social environment. This environment includes a value system (good, fast, cheap - pick two), norms of behavior (e.g., coding standards, design methods), and a social hierarchy (e.g., project organization, interpersonal relationships between team members). After sufficient experience looking at development from this perspective, we may find ourselves inclined to codify these structural characteristics into "approaches" (although we may not be able to articulate all of the interrelationships and interdependencies). We might even construct software development life cycle models and standards.

At higher levels of abstraction, the structural model looks at behavior and agents in the same way the coarse grained strategy does, but unlike its coarse grained peer the structural strategy can be applied to the psychological realm and to personal experience. Patterns and processes are its forte: this model doesn't need genetics, population biology, or actuarial data to generate software development life cycles, just a reasonable sample of development projects to examine and some basic understanding of how people are motivated or demotivated. If the results of coarse grained analyses are generally intended to be predictive, the results of structural investigations are pragmatic. A structural analysis of product deficiencies generates standards and tools, but it does not ensure a quality product. Similarly, a structural analysis of language generates lexicons and grammars, not poems.

THE EXPERIENTIAL STRATEGY

When we look back over our own personal histories at the people we've known or at the things we've done well or badly, our field of view is strewn with relics from all the experiences we've had. Ten years after the fact, I probably won't recall the quality of the lighting in the restaurant where I had dinner with a friend. But I will remember that he convinced me to sell my house to his sister at a tidy profit and that allowed me to realize my dream of going off to become a Zen Master. Our memories are partial representations of our individual pasts, partial because they are incomplete and partial in a quite different sense because they have been filtered by who we were at the time and who we are today. Like any archeological relic that tells a story to those who can 'read' it, our memories still 'speak' to us despite their incompleteness. They remain effective precisely because they are partial in the second sense above and reflect back to us the person we, today, think we were.

Clearly, this is not the terrain on which the empiricist is comfortable. The laws of scientific explanation do not hold sway here. Newtonian mechanics is filed under the category of 'memories from school', it is not a paradigm through which we view our pasts. This is the realm of images and language and 'felt experience'; there is room here for imagination and intuition as well as logical and mathematical cognition. What I am calling the experiential strategy has been applied in different forms and under different names to cultural, social, and psychological behavior by other disciplines (e.g., cognitive psychology, hermeneutics, and deconstruction). Similar to the 'historical narrative' strategy of Harvey and Reed, it is intended to discover 'purpose' not patterns, the 'meaning' not the mechanics of an event. We will find this strategy salubrious as we examine some software systems in the next section.

But before we combine systems with strategies, I should mention a particular application of the experiential strategy that will figure prominently in a later chapter: the linguistic analysis of experience. Linguistic activity underpins most activity that is not viewed as purely materialistic. We can (and in chapters 6 and 7 will) examine code production as a kind of linguistic activity that is subject to the same influences as normal speaking or writing. Coded "statements" processed by the CPU may lack a speaker, but they do not lack syntax, semantics, purpose, or effect. As is the case with the "historical narrative strategy" Harvey and Reed identified, this particular analysis method has been used infrequently (Singh 1992, Kelsey 1996) and has not had a chance to demonstrate its usefulness. But as we shall see in chapter 6, it is fundamental to our understanding of Complexity in software.

SYSTEMS, STRATEGIES, AND LIMITATIONS

We've spent so much time on the issues of spheres of influence, systems, and strategies because these are the fundamental methodological building blocks of any analysis of Chaos in software. If one takes the coarse grained view, some aspects of software are accessible and others are not, and the strategy also limits what sociological, psychological, and linguistic factors are considered. On the other hand, the experiential strategy has little predictive value. If one defines the system too narrowly or uses a limiting strategy, it is possible to generate misleading or invalid data about the system. And since developing and using software are both intentional acts, cultural, social, and psychological influences are hard to ignore. While these influences are extremely difficult to specify, they are crucial to understanding the potential for Chaotic behavior in software. On any strategy other than the coarse grained *they are part of the initial conditions of the system.*

In the first section of this chapter I listed a number of questions we must answer before we can call any system Chaotic. Abbreviated, here they are again:

1. Are the key characteristics of Chaos present?
2. Are we examining a system?
3. Can we justify our claim about Chaotic behavior mathematically?
4. Can we justify our claim about Chaotic behavior formally by appealing to other relevant disciplines?

5. Is it reasonable to assume that if we cannot meet condition 3 or 4 today we may be able to do so in the future?

Question 1 requires us to probe characteristics of Chaos we have not yet discussed, so we must defer an answer until the next chapter. At this point, I believe we can answer yes to question 2. The phases and tasks of a development project are discrete entities on the project plan but in execution they are interrelated and interdependent; the human resources are line items on the expense sheet but a team in practice. The developer works within the constraints of the execution platform, nurtured by his or her own experience and savvy, and guided by the capabilities of the programming language being used. None of these systems are mere collections. As for questions 3 through 5, they receive different answers depending on the system and on the strategy, and we can now illustrate this with a few examples using our previously identified systems.

PURPOSEFUL CODE EXECUTION 1 – TESTING

For the sake of argument, suppose the tester is testing the software for compliance with some typical software quality factors. (For the purposes of this illustration we'll ignore tester capability, level of alertness, and other such psychological and physiological variants.) We'll use the quality factors of Reliability, Correctness, Integrity, and Testability from the set of factors presented in Evans, M. W. 1987, Perry 1987, and reproduced in Pressman 1997. (One could as easily use a host of other variations on the quality factor theme. Some possible strategy/quality factor combinations are given in Figure 4.4.) These factors lend themselves to coarse grained analyses. Reliability is calculated by treating the program as a black box. Correctness compares such coarse characteristics as requirements and outputs, etc. These quality factors also tend to be applied to program or product characteristics that have binary values or at least a small range or variance. The security features of the product do or do not allow break ins, file write collisions, etc. The product does or does not fail in some period of time. The product can or cannot be tested against the requirements set.

For the tester, the "object of study" is deliberately coarse grained, and the study, the analysis, is undertaken for the purpose of validation and verification. If Chaos is to be found in this system, the-execution-of-the-program-in-a-test-environment system, we are well positioned to model that system mathematically.

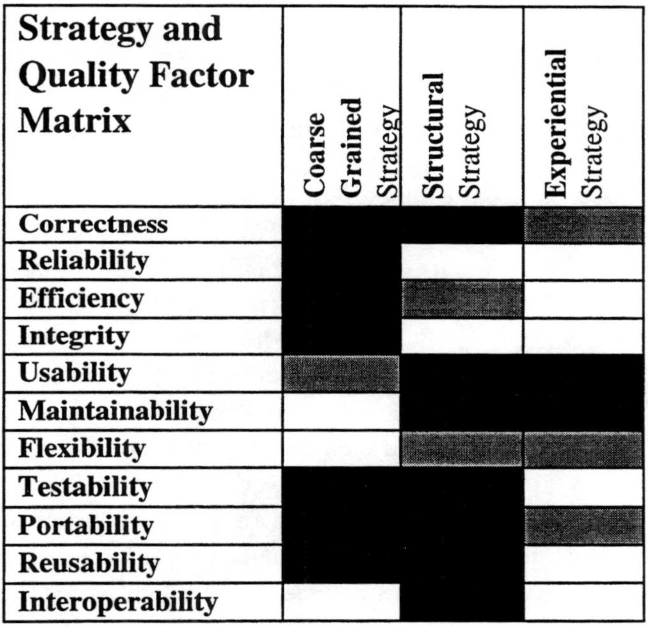

Strategy and Quality Factor Matrix	Coarse Grained Strategy	Structural Strategy	Experiential Strategy
Correctness	■		▓
Reliability	■	■	
Efficiency	■	▓	
Integrity	■	■	
Usability	▓	■	■
Maintainability		■	■
Flexibility		▓	▓
Testability	■	■	
Portability	■	■	▓
Reusability	■	■	
Interoperability		■	

Figure 4.4 - Applicability of strategies to quality factor analyses.
In this list of quality factors (taken from Evans 1987 and Perry 1987), dark shaded areas represent strong affinities between the strategy and the quality factor. Lighter shaded areas indicate weaker affinities. Martin 1983 contains lists of characteristics that can be examined from a coarse grained view; Möller 1993 lists coarse grain metrics.

Looking over the incidence rates of defects across the duration of the testing effort, we might find that the incidence is somehow cyclic or reflects a certain type of stimulation. Suppose defect rates increase when certain types of tests are executed, or defects appear in certain functions and not in others. We could (without recourse to sociological, psychological, or linguistic data or hypotheses) derive some relationship between defect incidence and test type, or between defect incidence and operation, or between defect incidence and the 'locations' stimulated in the code.

If we could then apply some Chaos model to determine that there was a nonlinear but also mathematically noncoincidental relationship between the number of defects and test type, operation, or location, we would have a very good case for claiming that this program under these test conditions exhibited Chaotic behavior. We would not however be prepared to extrapolate that conclusion to other test efforts – we have only a coarse grained analysis of a single instance. Our data is not only insufficient in quantity to support any extrapolation, it may also be insufficient in quality.

The coarse grained strategy is so useful because it works with abstractions: in this example we used it to look at defect rates irrespective of other defect characteristics. Its greatest strength is also a potential weakness. If we ignore information about the system that is available to the structural and experiential strategies, we may not see the limitations inherent in our analysis. When we count defects in the category Correctness discovered during Systems Test, we are not looking at the relationships between the requirements, the developers, and the product. The requirements may not have been clear, even after they were reviewed. The developers, as they are sometimes wont, may have decided that they knew the customer's needs better than the customers did. Design decisions may have caused some requirements to be only partially implemented in the program. A host of other reasons can be cited, but they will invariably come from our understanding of how customers and developers behave and think and how the project has been arranged structurally. Without probing the context for the defect (often referred to 'classically' as the "root cause"), we cannot move from this instance of Chaotic behavior to any other instance of software development.

There is an irony here we should not miss. Our investigation may have succeeded in justifying mathematically the claim of Chaotic behavior in this instance. The same "approach" may be used to determine that Chaotic behavior appears in other software test efforts. But the significance of that finding for software development in general is severely limited unless we can also justify it formally by appealing to experiential and structural factors. Otherwise, the multiple incidences of Chaotic behavior are possibly coincidental; they are not 'grounded' in the larger system of software development. To use phrasing from chapter 2, the investigation approach may not be able to claim either accuracy or precision. It has not shown that the factors being considered in defect incidence are necessary and sufficient, and it has not shown that the effects of these factors are independent of other factors the approach does not consider. This is why software must articulate its assumptions underlying, and the complete context required by, its "approaches".

PURPOSEFUL CODE EXECUTION 2 – GETTING SOMETHING DONE

Let's point our spyglass 'downstream' a bit. Product testing has successfully completed; the product has been shipped. Now a user has to execute the program to accomplish some task. In this context, the program is incidental, it's merely a tool to get something else done. The software quality factors of Usability and Efficiency in this system, the-execution-of-the-program-in-a-user-environment system, are very likely to be qualitative. There is no escaping the sociological and psychological aspects when we examine a system such as this and metaphoric analyses are more likely than mathematical ones. In the previous example we could claim that defects and stimulations of defects (tests) were the important initial conditions: varying the latter caused an exponential increase in the former. Here we have few such "things" as defects and tests to work with. There's the user's processor – its performance speed and capacity can be quantified. But as for other 'initial conditions', they are intangibles like the user's expectations about performance, what 'easy to use' *feels* like, etc.

It would be tempting to say that one can't predict the user's reaction to the product, that there are so many variables involved that "it is simply a Chaotic system". From that premise, we might suggest ways of 'controlling' the user's response, such as implementing detailed Usability studies before shipment, implementing usability requirements reviews, or implementing Efficiency checklists that examine attributes of the code irrespective of the end user's psychology and physiology (e.g., Martin 1983). It's also tempting to hypothesize that "turbulence" in this system is caused when the first failure starts a chain reaction of psychological events in the user that eventually ends in the rejection of the product as bad or useless. How convenient if future Help Desk staffers could handle irate customers with a response like "The product screens may be a little hard to navigate through, but to call it a piece of junk, well, there's no reason to go nonlinear over it."

But we know that Chaos is a discipline, not an excuse. If we don't know (mathematically or formally) the relationship between components in the system, that is, if we don't know how one initial condition is affected by another, we also don't know how the system behaves or how to control it. My suspicion is that unless (or until) the social sciences can show there are Chaotic features in how people respond to the frustration of their efforts, software professionals had best be cautious about calling such systems Chaotic.

The Development Project

If the previous examples identified some limitations to the coarse grained strategy and the dangers of an uncritical use of metaphors, this example will try to present both in a better light. Indeed, here we'll examine some of the pitfalls of the structural strategy that can grow out of a perfectly reasonable use of the coarse grained strategy. The development-project-as-system differs from the others just discussed because it evolves over time in a way the test effort or the end-user's experience does not and because it is inherently more complex (I did not say Complex) than the others. We'll soon see that this evolution over time is a mixed blessing: while it supports coarse grained analyses it also turns individual project dynamics into a limitation on any general conclusions we might make and on any normative procedures we might devise.

If we apply the coarse grained strategy to a project, we'll abstract from its experiential context certain attributes and trace their changing values over time. We'll look at raw defect incidence rates, phase incidence and containment (Möller 1993, Kan 1995), reuse, productivity, workflow efficiency (Perry 1981), etc. We can compare the defect and productivity measurements with the project-level measurements of cost and time. Suppose our analyses show that defect incidence is related Chaotically to cost overruns (where there is no contingency budget) and/or variances from estimates (where there is). We know from the first example that our findings may be true for this project but not for others. We also know that to extrapolate from this instance to other projects we'll need to know more about the story behind the headlines on the monthly status report.

Since we know how projects are 'structured', we also know that defects have ripple effects (call it 'turbulence' if you are not prone to understatement) in a project. They cause other delays and conflicts. Time and resources committed in the plan to other tasks are diverted to diagnostic meetings, to fixing the problem, and to retesting it. Hardware may need to be appropriated from other parts of the project to do diagnosis or testing; people may also need to be appropriated. Bug fixing takes time away from feature development, bug testing stalls product function testing. Parts of the project can come to a complete stop, and as the Gantt chart lines get longer the costs increase, the corporate pressure increases, etc.

Variances from estimates actually look like a vicious cycle (quantified they might appear to be a limit cycle or an oscillation cycle). Assuming the project end date does not change and that additional resources are not added (a typical situation in IT/IS organizations), the variance increases the pressure to produce more in less time. That pressure increases the likelihood of mistakes due to exhaustion, loss of focus, etc. Mistakes further affect productivity, which results in more variances and more pressure, etc. On careful inspection, we may well find that defects can produce indeterministic nonlinear changes in other project 'variables' such as cost or estimates. Most of the factors in this vicious cycle are structural, that is, they have to do with the flow of the project timeline, resource leveling (regardless of the actual resources), budget, etc. Only the productivity 'variable' is psychologically qualified, and even that can be treated a coarse grained variable. So we seem to be in a position to extrapolate from one project to another on the assumption that the actual values will differ but the mathematical model won't. Let's assume we can.

But now what? The structural strategy is pragmatic. It assumes that there is a solution to the problem, that there is corrective (or preventive) action to be taken. Any seasoned software professional reading this example has already come up with half a dozen techniques for lessening the risks of defects with their deleterious effects on end of year bonuses. Let's look at one – code reviews.

There is ample evidence that code reviews (or inspections) are effective at discovering defects in the program. It would seem to follow that reviews would be a good structural addition to projects: they catch defects upstream, so they are a reasonable preventive action against cost overruns and variances downstream. If defects late in the project cause Chaotic behavior in projects, then any way defects can be removed earlier will limit the project's risk. It's a perfectly reasonable approach to the problem: we should be able to 'control' Chaos by adding contingency budget and by limiting the range of values for one initial condition, call it "latent downstream defects". If the defect rate never reaches a threshold value above what the contingency budget can handle, the project will not be forced into a far-from-equilibrium state where it starts a cycle of Change Requests and cost overruns.

It may be a reasonable approach, but it may not be logically sound. It may not even be structurally sound. And in that respect it once again illustrates the importance of critically reviewing "approaches" to software, not only for their ability to deal with 'classical' software systems but also for their ability to handle potentially Chaotic systems.

It is not the removal of defects upstream, but the *effect* of defects downstream that matters. Forty different instances of noncompliance with Windows programming standards may not have the same effect on the project as five different program exceptions that result in five different cases of data corruption. We certainly could perform multi-project analyses to determine that "reviews in this set of projects" decrease "the exposure to variances in this set of projects". But we are still at the coarse grained level of analysis. Unless we can specify definitions of "exposure" and "variances" that are project independent and yet also directly relate to a specific project, we have not made a pragmatically significant claim about the effectiveness of our preventive action.

The relationship between defects removed, exposure, and variances could well be coincidental. The key component in the allegedly Chaotic system is "exposure" insofar as that describes the dynamical effect a discovered defect has on the project. But the actual events of "exposure" are unique to each project; they occur in, are constrained by, and coeffect the social, psychological, and structural context of each project. Just as in the first example the incidence of Correctness defects needed to be supplemented with a better understanding of the context in which they were generated, so here the effectiveness of reviews at preventing downstream *dynamics*, not defects, needs to be demonstrated. And *that* requires knowing what these dynamics are and how well a review removes defects with deleterious dynamical effects. It has nothing to do with the defect removal rate *per se*.

Not all reviews are created equal. The experiential strategy must be employed to identify the prerequisites for "successful" reviews, where "success" is defined not as the number of defects discovered but as the number of *relevant* defects discovered. That strategy would be employed to discover maximal attributes of the review, such as the conditions under which it would be successful. These attributes might include the experience of the developers participating in the reviews (their development experience, knowledge of the platform and product type, etc.), the maximum amount of code that could be effectively reviewed (affected by reviewer concentration/attention span and the complexity of the code being reviewed), and so forth.

The industry already has much of this information (e.g., Dobbins 1987, Gilb 1995, Pressman 1997, to name a few). But the maximal attributes of the review would also need to be supplemented with a defect typology that reveals the impact of the defect, its effect on the customer, its penetration in the customer base, and its location in the program.[22] These features of the defect would in turn cast light on what it would take to fix, and test, a defect. And *that* information, discovered in the review but not a direct result of performing reviews *per se*, would provide the information needed to correlate defects discovered in reviews with a decrease in exposure to Chaotic behavior downstream.

SUMMARY

We've covered a lot of ground in this chapter. We've discovered many components of software systems and we've touched on the complicated interrelationships between forces that drive those systems. The task of analyzing software for Chaos may seem daunting at this point – where would be begin? Before we move on to examine system characteristics that are likely to reward Chaos-based analysis, let's collect the results of the investigation thus far.

- There are two means to justify claims that a system is Chaotic – the explicitly mathematical, and the explicitly formal.

- There are multiple spheres of influence both in a development effort and in our understanding of it.
- We have identified a number of systems in "software"; these are more than collections of parts whose only relationship is they appear on the same project plan or occur at about the same time.
- There are several strategies for analyzing systems. These strategies may overlook some influences and/or may not be suitable for all objects of study.
- We have discovered that the coarse grained strategy may provide data to mathematically verify Chaotic behavior although it may do so by ignoring other important aspects of the "object" being studied, thereby limiting its applicability elsewhere.
    - E.g., the strategy may mask out other possibly Chaotic behavior – problem solving may be chaotic (Torre 1995) but the coarse grained strategy looks only at the results of decisions not the decision process, at defects not their contexts.
- But we have also discovered that claims of "Chaotic behavior" can be model based or metaphor based, and that there are limitations to metaphoric applications. Such applications may identify non-quantifiable, structural components of Chaos that cannot be modeled. But these metaphoric uses of Chaos may not be transferable to other situations.
    - E.g., problem solving may be Chaotic by virtue of the metaphoric use of such terms as attractors and turbulence. As the person 'works through' the issue he or she is pulled first one way then another. But that doesn't necessarily help us solve problems that occur in project review meetings. What may be suggestive on the experiential strategy may be useless on the structural strategy.
- We have become more aware of the difficulties in controlling Chaos, especially when our understanding of Chaotic behavior is incomplete or we have focused our attention on the wrong attribute of the system.
    - E.g., the structural strategy may lead us to add components based on their apparent success in some circumstances, and the coarse

grained strategy is prone to extrapolation to other environments. But success in the new environment is not ensured unless the context of the original and the new application are specified and are similar.

- And we have outlined some of the methodological shortcomings of applying "approaches" indiscriminately, especially when examining Chaotic systems.

Chapter 5:
A Conceptual Model of Chaos in Executed Programs

> We may blunder in various ways in metaphysics without any fear of being detected in falsehood.
> Immanuel Kant, *Prolegomena to Any Future Metaphysics*, §52b

Taken out of context, Kant's remark may seem a bit snide even to those with a healthy distrust of top-down, one size fits all hypotheses (to say nothing of directives from upper management). In fact, Kant was exploring the relationships between different kinds of knowledge. He was interested in the capabilities and limitations of reason and of sense experience, and their dependence upon certain fundamental unifying 'principles' of thought. We need not plunge into the details of his philosophy here. By now we should be sympathetic towards his agenda, and we should have some sense of the difficulties he faced. For our purposes here, his remark serves as a reminder that a top-down approach is only as good as the assumptions and data behind it.

Limitations to a Top-Down Effort

In the previous chapter we spoke of nested systems, starting with the code executing on the processor and ending with the project that created the software package. We can of course take an even broader perspective. We could consider the production of software or its purposeful execution as instances of sociological and psychological structures. From that perspective, many of the tasks and decisions and deliverables in the software development life cycle could be reduced to more abstract sociological entities. Code development would be a specific instance of 'task fulfillment'. Requirements analysis would be an instance of negotiation. Development could be an instance of 'worker productivity'. I'm personally not convinced that reducing software to sociology and psychology will in fact simplify things for us – the devil might still be in the details. In any event, it is not an immediate solution to the problem.

From psychology we will need to know if there are Chaotic characteristics in people's behavior as it relates to software development and evaluation. For example, how do people (developers) handle the frustration of their efforts to produce something? How do people (users) respond to the frustration of their expectations? This information would shed light on how developers deal (or don't) with requirements drift and schedule changes, or on how users respond to a product that doesn't 'measure up'. We also will need help from psychology to understand the dynamics of negotiations. This information would help us understand how Chaos figures in reviews. Where lies the 'bifurcation point' in a requirements or design review: is it when the discussion is stalemated because a compromise can't be found? When upper management reviews a project and decides to cut functions to save costs, what are the "initial conditions" of the discussion around the conference table and how do they evolve? And we surely can use help in understanding the possibly Chaotic characteristics of problem solving. How does an individual work through a design trade off problem? How does a team do it?

Some of these questions have been addressed from a Chaos perspective, but there are certainly no universally accepted models or laws the software industry could appropriate today. Nor is sociology ready to provide us with the information we need. It would be extremely helpful if the software industry knew whether there were Chaotic characteristics in the effect of new technologies on IS/IT project performance. What about joint on-shore/off-shore development: is there any evidence of indeterministic effects on communication, productivity, or product quality? How does organizational structure in the software industry affect actual project performance, worker productivity, or the ability of the development project to recover from changes in plan?

So the top-down search for Chaos in software faces three significant challenges. First, the number of components and influences in software – whether by that we mean its production or its execution – seems unmanageably large. Second, in most cases the information we need simply isn't available. Even if we did dissect, say, a requirements review into all its constituent parts we would still need theories and models from the social sciences to determine where Chaotic behavior *might be latent* in that review. Finally, we don't yet understand how nested systems affect one another. If we did find that psychological or sociological Chaotic systems were latent in the requirements review, we would still need to determine to what extent such systems *might influence* the review.

Still, using a 'top-down' approach, from the most inclusive of the nested systems to the lesser inclusive, will certainly give the software industry a sound, formal foundation for its subsequent analyses of Chaotic behavior in the industry and in its products. Hopefully this book will stimulate research into those areas requisite to a detailed top-down description of Chaos in software. But in the interim we can still make some progress. It is important progress, nonetheless, because the top-down strategy has no access to one critical system, the execution of the program in the execution environment. That is not accessible to (traditional) sociological or psychological research, yet it is fundamental to knowing how software programs behave.[23] And of course knowing how software programs behave is instrumental in determining how users of such programs will behave, since it sets some of the limits on *that* system. So while we await the arrival of the "one and indivisible" Grand Unifying Theory of Chaos in All Human Endeavors, I'll ask for your indulgence as we look at a few more "little parts" of the puzzle.

BOUNDS, INITIAL CONDITIONS, AND TURBULENCE

These three concepts will prove to be problematic for any Chaos-oriented analysis of software. Except for a very coarse grained analysis of defects, costs, and other such 'objective' data, definitions of "bounds" and "turbulence" will in most cases not have the same rigor one would expect in a Chaotic analysis of a physical system. When software professionals speak of bounded systems or of turbulence, they are already speaking metaphorically from the perspective of the hard sciences. So the best we can hope for is to define what these terms mean within the specific systems we are analyzing.

Similarly, any specification of "initial conditions" will seldom have the precision of, say, a list of conditions such as momentum and mass. Where the coarse grained strategy is appropriately employed there is a better chance of specifying initial conditions than in those cases where the experiential strategy figures prominently in the analysis. Unfortunately, organizational structure, 'development culture', and the cognitive and emotional contexts of the people involved are clearly conditions of the development project, but they are either too gangly or too ethereal to fit neatly into state descriptors in an equation. Further, the conditions will vary depending upon what system one is analyzing and how one is analyzing it.

The situation is worsened because of the nesting of systems in "software". Something may be a boundary of one system, an initial condition in another. For example, the developer's facility with the programming language is a boundary for the system of the-code-being-developed-by-this-developer because it limits what the developer is capable of coding. But the developer's facility with the language is an initial condition for the development project as a whole. Even what appear to be "physical" conditions change their strategic value depending on the system being analyzed.

If we look at the code executing in the processing environment, the hardware's capabilities present a specific boundary to our analysis. If instead we are interested in the system of project development, hardware boundaries during execution are an initial condition. For example, from the project perspective it is not the execution on the processor *per se* that interests us but rather the successful completion of the test phase, which includes far more factors than what hardware is used. What is true of initial conditions is also true of turbulence. Taken to be a change of state or value, turbulence is meaningful only within a set of possible states or values, which in turn depend upon some notion of boundaries, be these physical, metaphoric, or system-specific.

To illustrate, let's look at two possible uses of these concepts. The first is a purely metaphoric application. The second shows how we might define these concepts for a specific system.

For the purposes of illustration, let's assume that arguments over requirements in a requirements review are 'turbulence'. We'll set the boundaries of the system under study at the beginning and end of the review, and the initial conditions we'll define as positions each side takes up as it initially sits down at the conference table. On this view of the system, "turbulence" ensues when the participants 'go back and forth' trying to come to a compromise that they never reach. I suspect many readers will find this application of Chaos Theory a bit shallow. While I would agree, I should point out that it appears to be a legitimate metaphoric application. If the software industry accepts any metaphoric analyses *at all*, then I'm not sure it can distinguish between 'good' and 'bad' metaphoric uses without moving perilously close to aesthetics.

On the other hand, we already understand nested systems well enough to help us define more precisely what we mean by bounds, initial conditions, and turbulence for any given system. While such definitions might never approach those used in physics, they would at least be structurally sound because they were developed from the system. We'll see this method at work in the rest of this chapter, but first let's finish the comparison by re-examining the requirements review.

Setting the boundary of the 'review' at the clock start and stop times is still permissible, but the bounds of the system must also include the 'social' pressure on the parties involved. Each side has a 'position' that it implicitly or explicitly takes up. The customer may not be able or ready or willing to negotiate out functions in order to stay in budget; perhaps the development team may not be capable of delivering the full requirements list on time at budget. These positions *function* as boundaries – this is not just a case of metaphoric usage. My "position", a complex collection of linguistic and psychological and social influences, circumscribes the kinds of things I'm likely to say in the review and the kinds of statements I am likely to accept or to challenge.

The initial conditions of the review include a myriad of physical, psychological, and social factors. Suppose the customer representatives flew in from the East coast overnight and slept little, and now they have to wrangle with an ornery but well rested and native development team at 7000 feet elevation in the Colorado Rockies. It's a fair bet that the initial physiological conditions are not in the representatives' favor. Fortunately, I don't think an analysis would have to consider every possible physical, psychological, and social factor. The system we are analyzing has some salient characteristics that we can use to help limit the number of conditions we have to monitor. If for the sake of illustration we assume that the goal of the review is to come to some agreement about the product and the project costs, then initial conditions become the actual requirements and the logistical assumptions about cost, difficulty, etc. on each 'side' of the dispute. During the course of the review, turbulence is possible when either:

- The two sides cannot agree on a common value for cost, difficulty, etc. - that is, when the end values for the initial conditions remain divergent,
- And/or when the review team reaches a consensus about the value of initial conditions, say, the costs of implementing the requirements, but these end values conflict with the position taken by one side or the other with respect to the *other* initial conditions.

This has been merely an illustration of how one could create system-specific definitions of some key Chaotic characteristics. It is certainly not a complete description of the review meeting considered as a system. In the rest of this chapter we will focus on one particular system, the software in execution, in order to provide as comprehensive a description as possible.

CHAOS IN EXECUTION: SOFTWARE-SPECIFIC DEFINITIONS

The concepts of bounds, initial conditions, and turbulence are really significant for us only in *Chaotic* systems. Let us ignore the less manageable 'systems' of the development project and the purposeful use of software because we lack the required information from the social sciences. We still face a number of challenges when we examine software in execution. First, we have to define what Chaos is for the non-physical system of software execution, so we can then identify what systems will be interesting and what systems won't be. Second, we have to use that definition to develop a list of critical attributes of the software and processing environment that may produce Chaotic behavior. Third, we must show how these attributes correspond to bounds and initial conditions and how they may promote turbulence. Let's begin by defining the "system" of executed software.

5.1) The executed software system is defined as the execution, in a specific environment, of encoded components (statements, directives, and protocols) for a specific purpose with anticipated operational outcomes.[24]

"Environment" includes one or more (usually more) processing agents, the code (language used, compiled executable, etc.), and the code paths and the protocols that are activated *in real time*.

"Purpose" can have two meanings: the extrinsic reason the software execution is invoked, and the intrinsic, computational goal of the software's actual execution. A user may open a word processor to create a letter, and this extrinsic purpose is intrinsically supported by the execution of appropriate code paths that create the form, echo and store the input, etc. In what follows we will not need to distinguish between these two senses, but it is important to point out that for any instance of execution, the purpose is one of the bounds of the system.

"Anticipated operational outcomes" is more problematic. In fact it lies at the heart of the issue of what Chaos is in software execution. I suggested some time ago that when we take tenets of physical Chaos and apply them beyond the confines of physics we must break new ground. We're now ready to do just that.

A physical system can be called Chaotic if it exhibits behavior that can be mathematically modeled using a model that the scientific community agrees is a description of Chaotic behavior. Both aspects – the conceptual model and the equation that describes it – are critical for the scientific application of Chaos. But we do not have a mathematical model for software execution – at least not for the complex systems we are interested in. We will have to rely on structural attributes of systems, rather than their conformance to mathematical models to distinguish Chaotic from non-Chaotic systems. And since software execution is by its very nature purposive, we can use anticipated outcomes as a means to distinguish Chaotic from non-Chaotic behavior – with some important caveats.[25]

Random behavior is not Chaotic behavior. If we created a program that could actually generate a number at random, whose algorithm gave us no means to predict (even stochastically) the outcome, we would not have a Chaotic software system.

Accidental behavior is not Chaotic. If a failure in one program produces a failure in another, that secondary failure may be unanticipated but it is not Chaotic unless the two programs are structurally linked.

For example, take the case of two desktops on a network exchanging data between them as part of some background job while their respective users work in some other application. During the data exchange, desktop A times out desktop B, and resends desktop B some request. Because of intermittent hardware failures on the physical bus between the desktops, the time out condition persists. Desktop A sends more duplicate requests to its peer. Desktop B has actually received all the requests sent to it, and although it sent a "My Brain is Full" message back to desktop A sometime ago, the requests just keep coming in. It now has dozens of requests on its queue where it might have had only one. The queue management policy was designed for 6 entries. While the queue overflows, say into protected system space, desktop B finishes the first request and sends the requested data to desktop A. The protocol between the nodes ends and B purges its queue. A few minutes later, a spreadsheet application running on B throws an exception and terminates.

The demise of the spreadsheet was certainly unanticipated. It could not have been predicted. But it is accidental, not Chaotic, behavior: the data exchange program and the spreadsheet are not structurally related.

On the other hand, suppose the data exchange program on desktop A had been written differently. Suppose on the first time-out it breaks its connection to desktop B, then tries to re-establish the connection and restart the data transfer. Under the right conditions (where the connection messages are lost on the bus but the data is not), desktop A may enter a processing loop. It terminates the connection, reopens it, and restarts the data transfer. Then the intermittent bus failure causes the connection to time out, and desktop A reiterates through the same steps. In this loop the data exchange program on desktop A is exhibiting Chaotic behavior. Instead of a logically linear execution path, the program is oscillating between states (shutting down connections, reopening connections, restarting operations). This is not the anticipated behavior, and the program is no longer achieving its purpose.[26]

Expected and purposeful behavior also allows us to define "equilibrium" and "far-from-equilibrium" for the system of executing software. In turn, that helps us define other terms such as "nonlinear".

5.2) The 'equilibrium state' of the product is achieved when the product operates in compliance with macrolevel and microlevel expectations.

For our purposes here, the macrolevel expectations are defined by the product specifications, documented or not. We know that for some *other* software systems, such as the software being used in a user acceptance test, a specific user's expectations may also be relevant macrolevel expectations. In this case, however, our limited definition is acceptable because it is just the execution that interests us. Also for our purposes here, the design, interface, and functional specifications for the product define the microlevel expectations for the product. Although we will find later on that the expectations of the developer are a critical component of software complexity and in some cases of Chaotic behavior, we will postpone discussing this factor until a later chapter.[27]

5.3) A far-from-equilibrium state occurs when the actual performance of the product differs from the expected performance or when the actual code paths executed veer from the expected code paths.

Not all far-from-equilibrium conditions result in Chaos. They may result in failures of any sort, Chaotic behavior being only one type of failure. Chaos ensues when the code paths take unanticipated trajectories from which the system cannot gracefully recover.

The use of the term "trajectory" from physical Chaos Theory is deliberate. That concept provides us software specific definitions of "nonlinear", "dynamical", and "turbulence".

5.4) Programs exhibit nonlinear behavior when they deviate from their intended 'logic' or 'flow', that is, when some microlevel expectations about behavior have been violated.

This nonlinearity is obviously not the same as the exponential change in the value of a state variable in a physical Chaotic system. But I don't believe that the use of "nonlinear" in this case is *just* a metaphor. Most of us think in a linear fashion; when developers "plan out" how software will behave they certainly think linearly. The use of the term "nonlinear" here is justified on experiential grounds, if not mathematical grounds. Software behavior becomes nonlinear in the experiential (or psychological, or phenomenological) sense when that behavior is *not* part of the sequence of behavior the developers thought they were creating, or the observers expect to see. Plotted out, it may look oscillatory as in figure 5.1, and may be one indicator of Chaotic behavior.

5.5) The system of executed software (which as we know includes more than just the program) exhibits dynamical behavior when nonlinear program behavior spawns systemic behavior that violates the purpose of the program and/or changes the anticipated outcomes of its operations.

Immediate program termination for whatever reason (say, divide by zero) is not dynamical: there is a precise 'termination' route that doesn't involve going back over code paths already invoked for the operation that's now terminating. The loop on desktop A *is* dynamical, since it is not a direct termination routine. It involves other parts of the program and it iterates (loops) over code paths already taken as part of the operation that is blocked. Note also that if the system includes more than one processing agent (definition 5.1) the *systemic* behavior can include the behavior of the software on the other agent(s).

5.6) The system's unsuccessful attempt to recover from this dynamical behavior is turbulence.

There may be levels of turbulence in any system. As desktop A loops through its recovery and restart operations, it is exhibiting turbulence. Simply illustrated, its code "flow" oscillates (see Figure 5.1). But if desktop B is also undergoing some recovery and restart iterations, the system itself (nodes and code and hardware environment) is also experiencing turbulence.

We now have a working definition of Chaos in "software", where "software" is taken as the executed software system. We also have transposed key characteristics of Chaotic systems, such as equilibrium and nonlinearity, into software-system-specific terms. Now we must work the reverse direction and look for attributes of software systems that might contribute to Chaotic behavior.

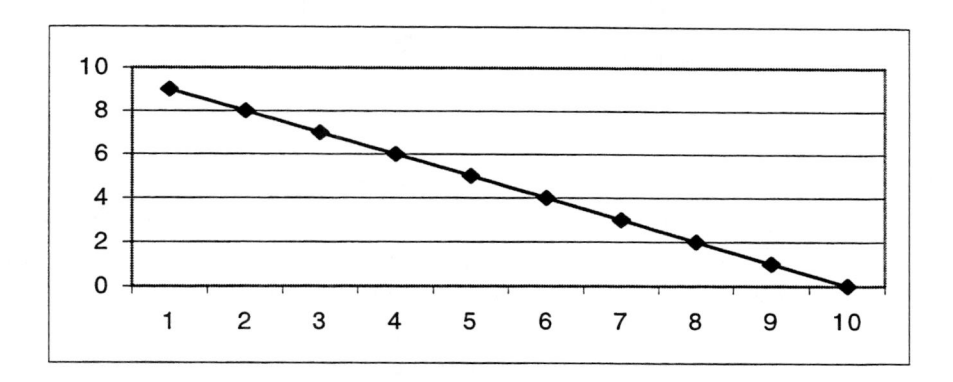

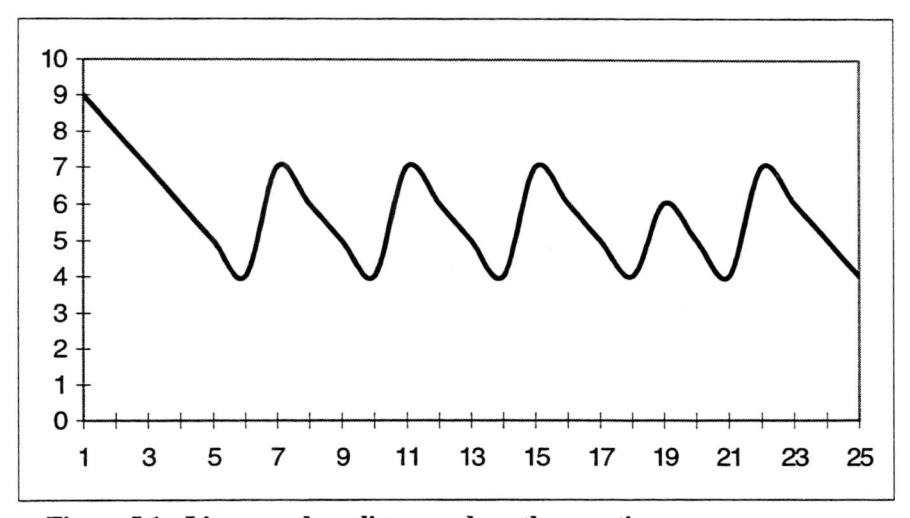

Figure 5.1 - Linear and nonlinear code path execution.
Plotting the logical or chronological execution of functions against time provides a way of distinguishing linear from oscillatory (possibly Chaotic) behavior. For a ten step operation, the top graph shows its sequential execution. The bottom graph shows it iterating over several interim steps due to some internal operational loop.

Attributes of Potentially Chaotic Software Programs

I think most readers will agree that some types of programs will not, *in execution*, exhibit Chaotic behavior. These programs have linear execution patterns in both normal and error paths, and their purposes and anticipated operational outcomes are limited or at least easy to specify correctly and exhaustively. Such programs might include the following: a standalone spreadsheet application, a supervisory shell or scheduler that merely removes entries from one queue and places them on another, a program that periodically reads data from some hardware measurement device (e.g., a thermostat), and simple forms of robotic control programs.

The salient features of these kinds of program are:

- They have no particular hardware affinity (the spreadsheet) or a very precisely specified hardware affinity (the robotics).
- Their functional logic may be complex but it is not "dialogic", that is, the program is in complete control of its environment, inputs, outputs, etc. (the scheduler).
- There is a limited range of supported operations, where an operation is something that is initiated by an external entity such as a user or another program (the supervisory shell).

  - The program accepts only limited feedback from its environment, and the feedback to the program is treated in a "linear" or single-threaded fashion (the thermostat measuring program or simple robotics).

What an irony that we should want to call programs that probably execute flawlessly uninteresting, but for our purposes here uninteresting they are!

Conversely, one might find Chaotic behavior where some or all of these attributes are absent. Put differently, Chaotic behavior may lie dormant in programs where the following conditions are true:

- There is imprecise, mutable, or mediated "hardware" affinity (where the "hardware" may in fact be a combination of hardware and

firmware or even hardware and controlling software, as in intelligent device controllers).
- Functional logic includes the "dialogic functions" of protocols or decision capability.
- The supported operations are many, broad in scope, or complicated in the number and interrelationship of tasks involved.

- The program supports distributed execution, is multithreaded, and threads contain multiple decision points that have chronological or sequential dependencies.

Let's look at each of these "conditions" in more detail, keeping in mind that we are looking at the potential for Chaotic behavior irrespective of time of execution or the reason for the behavior. There may be differing *reasons* for Chaotic behavior on ship date and Chaotic behavior three years later, human error in the first case and "spoilage" in the second, but at the moment we need to identify the *structural components* of that behavior. Only these can tell us if the software has the potential to exhibit Chaotic behavior because they are the crucial "initial conditions" of potentially Chaotic systems.

A physical Chaotic system shows "sensitive dependence on initial conditions" in the sense that the laws of the system are such that varying one condition produces nonlinear effects in the system at a later time. In a software system, these structural components function like the laws of the physical system: they "set up" the system to act Chaotically. This gives us a working definition of "sensitive dependence" and "initial conditions" for software execution:

5.7) Given a *specific instance* of one of the attributes of hardware affinity, dialogic functions, and operational complexity, and given some triggering series of events, the system's behavior can become Chaotic because of the way these attributes structure systemic behavior.

"HARDWARE" AFFINITY

Imprecision and mutability in the software/hardware interface can have several causes: concurrent development of hardware and software where the interface changes regularly, ambiguous or incorrect interface specification, changes to the hardware after the software has been released, just to name a few. Such events can certainly decrease the reliability or performance of the software product, but by themselves they cannot produce Chaotic behavior (definition 5.3). Chaos may ensue only if failures in the interface cause the program to iterate through code paths.

As described so far, hardware affinity isn't structural – it's merely the triggering event. There is however a form of hardware affinity that is structural: a hardware-software interface that is mediated by protocols and/or structures that separate yet connect, logically if not physically, the hardware from the software components. Such mediation occurs for example in message-based network or storage communication protocols or in queue based communication and command systems.

The command and status message protocols between a software device driver and some form of intelligent device controller represent, in part, the interface between the device driver and the device itself. But the interface is not immediate. It is instantiated in packets or messages that must be coded, sent, decoded. The purpose of the protocol, what it is trying to accomplish, provides a structure for both successful execution paths and for error handling. The communication carried on over such protocols is also usually chronologically dependent – information is expected to arrive in some sequence. It doesn't matter whether the protocol occurs at an architecturally high level on a network or in a simplistic packet queuing system over a bus internal to the processor. In the latter case, the timing dependencies of the network are replaced by timing dependencies introduced when the queue management operation is forced to store its current context across one or more interruptions.

It is of course possible to write software or hardware firmware that acts upon only a limited subset of a status-rich protocol. As an "initial condition", mediated "hardware" affinity *per se* is only loosely coupled to Chaotic behavior. The presence of a protocol or other communication mechanism is of less importance than how it is used and/or how it is stimulated. One or more of the attributes discussed below must also be present.

"Dialogic" Functions

The dialogic capacity of a system is the crucial condition for Chaotic behavior in the execution of a software program. A protocol or a messaging system has a syntax and a semantics; it contains "statements" that "mean" something, whether these statements take the form of state bits or data packets or status codes. It is a language, and therefore is at least potentially subject to all the vagaries and imprecisions of linguistic activity in general. We know from our analysis of Stephen Kellert's view of Transcendental Impossibility that reliance upon language may place this kind of software execution at risk from Chaos. But we should be able to probe a little deeper into the conditions in which this might occur.

Two tape recorders in playback mode across the room from each other is not a dialogue. Nor is a screaming match between two people. Dialogues occur when decisions are being made 'in real-time' based on the "input", or when there is a response to received information that is based on that information. For the dialogic condition to be met, then, the software must be able to respond to, not just acknowledge, all 'items' in that protocol, and it must 'make decisions' based on the content of the communication regardless of the 'other entity' it is communicating with. The dialogic functional condition (of which the mediated hardware affinity is a special case) provides a general chronological and sequential structure to program execution, and constrains program operations within its semantics and directives.

Operational Complexity

Communication protocols between nodes on a network are essentially dialogues between computational entities. If the software supporting this network has been properly architected, 'lower level' transfer mechanisms and protocols are unknown to 'higher level' programs. This separation provides flexibility and portability, and it also restricts the kinds of information that need to be passed from the local layer to its kin on a remote node.

Yet behind many communiqués between nodes there are a myriad of operations, probably at many 'levels' in the node. When a server returns a "task completed" message to a client, it's implicitly stating that all the resource management, computation, storage, and processing involved in the client's request has been completed. Not all of these activities were the responsibility of the program itself. Some were completed by the operating system, file system, device drivers, etc. This is the context behind the communication: the collection of operations that are, as it were, represented by the single status or state in the protocol called "Done".

Operational complexity adds a destabilizing factor to the dialogic function if it offers more opportunity for iterative behavior. It also may complicate the context, since operations require time and resources, both locally and remotely. If the protocol provides a general chronological and sequential structure, the operational complexity provides its specific instantiation (definition 5.7). The protocol doesn't change because of the contexts in which it is used, any more than the words in the dictionary change when used by different speakers. But the executed protocol, the 'internode conversation' as it were, is uniquely dependent upon the actual code paths executed on each node.

One important kind of operational complexity is the program's capacity to receive and to respond to multiple inputs from other entities and the complexity of the code paths taken in these responses. If the protocol and the operations are as simple as "Do this – *OK* – *Did it* –Thanks" there's little room for dynamic behavior. But if the same program on Node A is multithreaded, is engaged in several such operations and protocols, and has even moderately sophisticated error handling, the potential for Chaotic behavior on Node A is increased.

For example, programs that support internode synchronization in "clusters" or in multinode networks will be multithreaded, they will be able to handle multiple "inputs" from multiple locations, and they will be tightly coupled to the timing and recovery capabilities of the protocol(s) used to "work with" the other nodes in the network. In the case of synchronizing programs such as this, there are more code paths that can be followed and more functions that can be executed. There is an increased possibility of timing or sequencing anomalies. And there is competition within the program as its threads maintain several dialogues with other nodes at once, each of which may have widely divergent demands on the program's resources. For such programs, the context for any given code path becomes less of a dialogue and more of a round table discussion with a corresponding increase in the possibility of Chaotic behavior under the right conditions.

SUMMARY

Thus far we have transposed several of the central concepts of physical Chaos Theory into the system of software as executed in some particular environment. We have been able to provide system-based definitions, rather than use the terms from Chaos Theory in a metaphoric fashion. We have also identified certain attributes of software programs (and, thereby, of the system in execution) that function as structural components in a Chaotic system. While the presence of these attributes does not mean that a particular program or a particular system will in fact behave Chaotically, their presence does indicate a potential for Chaotic behavior. Stimulated by events that cause certain code paths to be executed, the program or system can perform unexpectedly while at the same time it performs within the structural constraints of the attributes of hardware affinity, dialogic functions and operational complexity. It appears that there is "order" even in software Chaos.

The move from physical Chaos to software Chaos was not seamless, however. We made a distinction between the unanticipated and the accidental, therewith introducing the first significant difference between Chaos in software and Chaos in physical systems: the role of intention and anticipation in software systems. In what has been called here "dialogic functions" we introduced a second major difference: purposive language use plays a far more significant role in software than in physical Chaos Theory. In introducing dialogue and anticipation, our analysis of Chaos has diverged from the trail taken by most other analyses of Chaos inside and outside of the physical sciences. Both dialogue and anticipation are intentional activities, and the notion of intentionality will help us in later chapters probe how the software code itself can produce, and structure, Chaotic and Complex behavior.[28]

For the purposes of this chapter, it was legitimate to ignore the "intentional" aspects of executing software: we chose to look at the executing system free (as much as possible) from the developers who created it. This allowed us to develop some structural attributes of potentially Chaotic systems. But if we are to fully understand Chaotic or Complex behavior in executing software, we *must* apply the experiential strategy. We will find that intentionality provides some of the structure for the executing system. It is thus the key to understanding nonlinearity and it helps explain why the attributes described above can circumscribe the purpose and the actions of the program and the system to create behavior that is Chaotic or Complex rather than random, ugly, irritating, etc. We will have more to say about these issues when we examine Complexity in chapter 7. To prepare for that analysis, we must first take a philosophical detour into experience, language, expression, and understanding.

CHAPTER 6:
COMPLEXITY AND THE ORDERING PRINCIPLES OF EXPERIENCE

> The wandering earth herself may be
> Only a sudden flaming word,
> In clanging space a moment heard,
> Troubling the endless reverie.
> W.B. Yeats, *"The Song of the Happy Shepherd"*, 18-21

Centuries of humankind's intellectual and social progress, reduced to clatter in a discordant and preoccupied cosmos ... no one ever accused Yeats of being 'up-beat'. But this view captures the alienation that troubled the late 19th and early 20th centuries. It was bad enough that science after Newton presented a purely material world unleavened by God's Grace and Purpose. But Freudian psychology and Darwinism threatened to reduce humanity to the status of libidinous bread mold feeding off the social structures and materialistic values installed by the industrial revolution. Those unaffected by the despair that filled Thomas Hardy's *Jude the Obscure* still had to face the First World War. Our collective conscience and faith in our own humanity shaken, we found ourselves orphaned from the very culture that had brought us to this impasse.

We would not soon recover, perhaps we haven't yet. Writing in mid-century, the quantum physicist Erwin Schrödinger wrestled with humankind's place in the mechanistic world he had helped to create:

> [The world presented by science is a] "... mechanical clockwork, which for all that science knows could go on just the same as it does,

without there being consciousness, will, endeavor, pain and delight and responsibility connected with it...." (Schrödinger 1996, 96)

If nature after Newton was no more than the "closed totality of the motions of spatio-temporally related point masses" (Kockelmans 1984, 216), where was there room for humanity's non-scientific achievements? Where indeed was there room for humanity?

[In scientific thought, man] "... takes himself – the subject of cognizance, the thing that says 'cogito ergo sum' – out of the world, removes himself from it into the position of an external observer, who does not himself belong to the party." (Schrödinger 1996, 93)

Schrödinger did not solve the problem in his 1948 Shearman Lectures. In our own day, Roger Penrose expresses a similar sentiment in his recent book *Shadows of the Mind*: "A scientific world-view which does not profoundly come to terms with the problem of conscious minds can have no serious pretensions of completeness" (Penrose 1996, 8). Humanity's place in the cosmos is still a matter of concern. We have already heard Sally Goerner's view of this "mystery" (Goerner 1995, 36). Finding a place for consciousness in the quantum universe is one goal of Henry Stapp's *Mind, Matter, and Quantum Mechanics*. And it is the main theme of Penrose's *Shadows of the Mind*. In this, perhaps ironically, such scientists and scientist/philosophers join the numerous writers in the humanistic disciplines who have argued that scientific methods and the revelations derived from those methods are limited by constraints intrinsic to the scientific perspective on the world.

ON THE BLEEDING EDGE OF COMPLEXITY

What does the mind-matter problem have to do with the possibility that some software systems are Complex? Why drag the philosophical debate about scientific limitations into a discussion of whether software may have emergent characteristics? The answer is: Complexity in software is best understood as a combination of the behavior of the product and the conditions of the product's creation. To understand Complexity we must leave the 'traditional' realms of science and explore thought and language. In order to provide a philosophical basis for Complexity, we must at least skirt the edges of issues of 'mind and matter', of perception and understanding. And where will all this lead us? In a few more than 25 words this is the thesis I will present in this and the next chapter:

6.1) The Complexity of a software system originates in the cognitive acts that create the software code; Complex behavior is stimulated by certain (possibly Chaotic) conditions in the execution environment; and the system is recognized to be Complex only if certain conditions (generally failure conditions) occur that force us to realize the software's actual behavior patterns.

It will take me some time to justify these claims. But if we look at the course of the argument so far, we can see that these claims are at least credible given what we know about methodology and what we have learned about the special challenges in analyzing software systems. In chapters 2 and 3 we explored the following points:

- The software profession needs to improve its understanding of software systems, and the software profession is well poised to do so in a more disciplined and 'scientific' fashion owing to:

  - Its current indebtedness to assumptions from classical science, and
  - Its desire to emulate a scientific discipline rather than an artistic tradition.

However, Chaos and Complexity present a new interpretation of "scientific method" that challenges fundamental assumptions that underlie the profession's extant tools and methods and behaviors.

- If the only relevant aspect of the new science were the mathematics of Chaos, the profession could be content to apply Chaos models to defect data, productivity data, and project success/failure data. It would only need one new tool to meet its goals.

- But taking the lead of some authors in the social sciences, we have learned that the Chaos models *are not* the only 'use' of the new science. We have identified multiple "systems" in software, many of which contain an experiential component that is not accessible to a coarse grained strategy.
- Further, we now understand that a methodological pluralism is necessary if we are to successfully probe and understand the variety of Chaotic or Complex systems.

  - The paradigm shift initiated by the new science does not swap systems for components, pieces for wholes, as the "object of scientific study". It fundamentally redefines "object" and "study".
  - Given the software industry's classical view of its own discipline, then, the new science forces the industry to retool rather than to hone the tools it has already.

But in the case of some classes of software system, such a 'retooling' takes us beyond the domains we are comfortable in. We saw in chapter 5 that we had to introduce intention in the form of anticipated or expected behavior in order to penetrate beneath the coarsest level of Chaotic behavior in software. That has several implications that will direct the course of argument in this chapter.

Software products are human artifacts: the clue to their Complexity lies not in the attributes of the product *but in its genesis, in the acts by which the product was created*. We cannot understand the act of creating software (or the act of using it) solely by using coarse grained strategies. And we cannot understand an emergent system by using linear, 'cause-oriented' tools. We cannot turn to positivist psychology, coarse grained sociological analyses, or other reductionist approaches whereby thought and behavior, especially the "linguistic" behavior of writing software, are viewed as causally determined symptoms of biophysical or sociological 'conditions'. That doesn't reveal the experience of writing or using software.

Furthermore, since we are interested in Complex systems, we are not interested in the details of a specific act of thought or a specific line of code. Instead, we want to learn something about the structural 'laws' of such systems. We will need to examine thought, imagining, and linguistic expression as *structures* of our cognitive behavior. In a very real sense these are the structural "initial conditions" through which the software product gains its potential for Chaotic or Complex behavior.

This, I fear, places both software and this book at the new science's "bleeding edge", as we like to say in software. Software systems are multifaceted. They include physical, cognitive, social, and epistemic components. One can't fully understand software systems without taking all of these components into consideration. Yet the new science has not progressed far enough to support the software industry in such a comprehensive analysis. In fact, as *an empirically motivated scientific endeavor* it may never be able to support that effort: not all structures are 'external' to us, not all structures are available to empirical analysis. Harvey and Reed recognized this, and made room for non-traditional strategies in their methodological pluralism.

As I try to explain how a Complex software system can come into existence, I will definitely be using a 'non-traditional' strategy. Indeed, I suspect that I will try the patience of some readers who believe that an analysis is scientific only if its results can be expressed in mathematical formulas. But we will in fact be probing behind the scientific attitude and into the structures of understanding that are prior to, and constitute, the acts of scientific understanding. To do this we will need to bring the discipline of phenomenology to bear on the problem of understanding. Phenomenology is a method for analyzing fundamental structures of our experiences. It shares with traditional science the goal of understanding why the world is as it is; to meet that goal it takes a radically different path than, say, physics and this has historically caused some friction between the disciplines.

For some readers, taking a phenomenological approach will be a paradigm shift in its own right and may add complexity to our analysis of Complexity in software. It may be best to contrast the 'scientific' with the 'phenomenological' perspectives in general before we delve into language, horizons, and projection. I should also point out that in what follows we will encounter a new twist on a problem that is typical of analyses of Complexity. While trying to describe the Complex aspects of teams, Ted Lumley points out that we are inclined to "force-fit linear-causal models" to what we understand, due especially to the fact that our language itself was developed to describe a causal world (Lumley 1997, 15). In our case, we will be trying to use language to show how experience and language are "ordering principles" of experience and language. They are not causes of phenomena, but structures or forms in which phenomena become knowable in the first place.

SCIENCE AS A STYLE OF THINKING

In "Nature and the Greeks", Erwin Schrödinger acknowledged the power of scientific analysis while at the same time accusing it of using "the greatly simplifying device of cutting our own personality out" (Schrödinger 1996, 96). By "personality" he didn't mean the quirks of character that make us prefer coffee to soda or a career in Information Technology to one in geophysics. He was drawing attention to the fact that our experience of the world, and the scientific descriptions of that world, seem disjoint. This is not a surprising position for someone who helped develop the quantum view of reality to take, nor is the tension between garden-variety experience and the revelations of science new with Schrödinger.

Indeed, that tension is, in its essential form, something we all live with everyday, but its effects aren't quite as striking as the difference between scientific laws and daily experiences. We only experience singularities – a dog, a house, a moment of conversation with someone we call a friend. Yet we "know" that this particular four legged, tail wagging, hand-licking entity is one of a class of objects we recognize as dogs. We "know" that the house is not just the façade we see, but is instead a three-dimensional dwelling with other sides that escape our sight but not our understanding. And we "know" that our interlocutor is not just a "talking head"; instead we understand the import of what she says because the conversation occurs in the context of our shared experiences and our "mutual" understanding of one another. Just as science abstracts from the manifold events available to it only those that will help it grasp the "laws" that govern some entity, so do we in our daily thinking move from sensory input to cognitive import, from merely undergoing an experience to understanding it.

However, it's the kind of abstraction peculiar to science that causes the greatest difficulty when it comes to finding a unified world-view that bridges the gap Schrödinger lamented in "Nature and the Greeks". Writing from the perspective of phenomenology, Hans-Georg Gadamer has suggested that "science" (before Chaos and Complexity, at least) has been successful only because it has been self-limiting. Insisting on methodological abstraction, it refuses to examine the external world in any other way (Gadamer 1977c, 11). "Any science is based upon the special nature of that which it has made its object through its methods of objectifying" (Gadamer 1977b, 93).

It is not that science objectively studies objects; rather, science defines its objects and then studies them. This may sound like a kind of circular argument – science proves what it sets out to prove. But we must be careful to distinguish methodology from circularity. Certainly, there are plenty of examples of what the philosopher of science and cognition Steven Horst calls the "fallacy of reduction" (Horst 1996): it is easy to abstract some features of an object for mathematical modelling and then assume that the abstracted features are all that really matter in the object.

But methodology isn't so much circular as it is "preconditioned". The history of scientific paradigms, the 'deconstructionist' movement in sociology and history of science, and even some Chaos proponents (e.g., Young 1995) offer ample evidence that such preconditioning has cultural components. But the effect that a cultural and historical context has on our understanding isn't the issue here – Gadamer and others would claim that such conditioning is an inescapable, and is in fact an enabling, condition of knowledge (Gadamer 1994, Kockelmans 1993, Feyerabend 1986, etc.). More germane is the methodological issue here, for that is where science and phenomenology often part company.

Phenomenology is a "style" of thinking that focuses on the "realm of immediate experience" and then moves from that realm to other realms such as physics and sociology (Kockelmans & Kisiel 1970, xii). This isn't mysticism: the structural components of perception are fundamental to the perception of a table, the ordering principles of understanding are fundamental to understanding "things" like laws, and the experiential structures underlying symbolic utterance are fundamental to the formulation of both sentences and mathematical laws. If you don't know how it is you perceive, think, and generalize, then you don't know the structure or the validity of your perceptions, understanding, and laws. Certainly, one can gain some confidence from such mental events if they seem to describe the world in such a way as to make it predictable. That in fact is one of the justifications of the "style" of thinking that is called "scientific" and contributes to what is called the "scientific worldview".

Recognizing these "styles" of thought and understanding how they affect us are the goals of phenomenology. If it does not take the positivist approach, it is because it cannot. Like the paradigm shift between seeing a two-tine and three-tine fork, one can not think both "styles" in the same instance. Only in the disparity between the views do the views become "views" rather than a naïve perception of a "thing". On the phenomenological view, scientific laws, mathematics, and language are all symbolic gestures by which we grasp meaning from the manifold of sensory inputs we receive from the "world" around us. In fact, they are the primary constituents of that "world", they make it an intelligible unity rather than a confusing morass of raw colors and shapes and noises (Kockelmans 1993 and 1966, Merleau-Ponty 1964a and 1962, Gadamer 1994 Part 3).

The scientific "approach" allows us to abstract certain features of that morass, to unify them, and to express that unification as an "object" that follows "laws". The objects of science are made into objects by appeals to some of the structures inherent in how we understand anything at all. Science finds objects within that 'horizon of understanding' that is its particular domain of expertise, and although it has created the object of study it assumes that the object is some how external and 'objective' (Kockelmans 1984, Szilasi 1970). In a twist on the age-old arguments over observation and theory versus belief, Wilhelm Szilasi claims that "natural scientific experience is theory" (Szilasi 1970).

As is, to some degree, all understanding. Theodore Kisiel, summarizing the views of the phenomenologist Maurice Merleau-Ponty, describes knowledge as a spectrum of styles and their objects:

"... the universe is no longer divided into neatly separated blocks isomorphic and parallel to each other but rather consists of a series of mutually inclusive 'fields' or 'worlds,' dominions that vary in dominance according to the phenomena under study and the perspectives employed." (Kisiel 1970, 261)

These "perspectives" are fundamental structures in how we understand the world around us. Hans-Georg Gadamer has explored how these perspectives affect our understanding of both history and of language. In his view, these preconditioning structures are habits of thinking. In translation, his term for this preconditioning is "prejudice" – an unfortunate choice for many readers –

but by "prejudice" he means an inclination to see things in one way rather than another.

> "It is not so much our judgments as it is our prejudices that constitute our being. This is a provocative formulation, for I am using it to restore to its rightful place a positive concept of prejudice that was driven out of our linguistic usage by the French and the English Enlightenment. It can be shown that the concept of prejudice did not originally have the meaning we have attached to it. Prejudices are not necessarily unjustified and erroneous, so that they inevitably distort the truth. In fact, the historicity of our existence entails that prejudices, in the literal sense of the word, constitute the initial directedness of our whole ability to experience. Prejudices are biases of our openness to the world. They are simply conditions whereby we experience something – whereby what we encounter says something to us." (Gadamer 1977c, 9)

By reducing "experience" to parts and things, science chooses a specific style of thought and works (quite successfully) within it. But the lawful structure of the physical world which science presents to us is, like other interpretive and analytical efforts, circumscribed by horizons intrinsic to the style: science's "physical laws arrive at an explanation not *of* the structure but *within* the structure which is its condition of possibility" (Kisiel 1970, 262). "We" are more than molecules and electrical impulses; consciousness is more than computational algorithms at work. That is not to say that neurobiology and the computational theory of mind have nothing to say about "us", just that they are one of several voices in the chorus.[29]

> "The facts only suggest, man gestures in a scientific way, and both movements belong to a single logic of gestation which lies at the heart of the genesis of any human meaning." (Kisiel 1970, 266)

Nothing that has been said so far detracts from the contributions scientific inquiry has made to our knowledge and to our cultures. On the contrary, such contributions have indeed revealed "truths" about the world. But they have done so not because they are methodologically impeccable. Science is not successful as science because an object-oriented view of the world is in some sense more True, or because it is a-personal or 'objective'. Instead, there is truth in science because it is one of the innate means we use to understand anything at all, part of a sentient tool kit that precedes scientific inquiry, informs it, and anchors it to the rest of experience.

Indeed, what has been said thus far from the phenomenological perspective is in some ways a formal justification for the methodological pluralism of Harvey and Reed, which we have described and employed in previous chapters. Of the methods available, scientific understanding is attractive for pragmatic reasons. Scientific laws are far more predictive than our personal hunches and guesses are, and the body of scientific information certainly seems to come closer to a cohesive body of "knowledge" than do our collective opinions and experiences. And that is because science reveals some of the "structures" of physical existence, of a mode of Being in general rather than of a thing in itself.

However, science may not be so effective at discerning structures of conscious existence without reducing expressions to accidents and the conscious to the physical. One can certainly look at the artifacts created by conscious beings – pictorial art, poems, cultures, software - as things *or* as expressions. We saw in the analysis of Chaos that the coarse grained strategy can help us understand some of the characteristics of Chaos in software. But to understand Complexity, we will need to shift to the experiential strategy and to look at the structures of perception, understanding, and expression that help bring Complexity into existence.

STRUCTURES IN EXPERIENCE - 1: INCARNATE EXPERIENCE

We already know that beneath the façade of macroscopic objects lies a puzzling world of microscopic components and interacting forces. We regularly stub our toes on table legs that, at the particle level, are anything but "solid". For those of us who are more at home with soccer than with String Theory, physical and mathematical theories remain an intellectual abstraction, despite the claim that they are "true" and somehow reflect the "reality" of what goes on in the world about us. Schrödinger describes this polarity this way:

Complexity and the Ordering Principles of Experience 113

"... we do not belong to this material world that science constructs for us. We are not in it, we are outside. We are only spectators. The reason why we believe we are in it, that we belong to the picture, is that our bodies are in the picture. Our bodies belong to it." (Schrödinger 1996, 95-96)

The "reality" we experience is not an abstraction, rather it is part of 'who and what we are'. It is incarnate not ideate, its 'reality' lies in its being experienced. It certainly *can* be the object of critical reflection; trends and rules may be discernible through analysis. But in the latter case, we are spectators. In other cases, we are players.

For example, consider the last time you raided the refrigerator while the world lay cloaked in darkness and your own mind was still clouded by sleep. You rose from bed and trundled down the hall to the kitchen and grabbed a, well, whatever it is that you content yourself with at 3 a.m. in the dark. So described, it seems simple enough. But it was in fact a highly complex, and revealing, set of actions. Unaided by your usual ocular triangulations, you navigated around the bureau's edge that you occasionally bump into when you're in a hurry. Without the aid of light and without leaving grimy fingerprints on the walls, you plied a reasonably straight course down the hallway. No map guided you to the refrigerator. Your hand effortlessly found the door handle, as if the handle had been placed exactly in your hand's path. This is the incarnate world, where objects and space are not things apart from your sighted, cogitating self. They are instead part of the fabric of your bodily movements. You and the world you experience are cut from the same cloth.

This incarnation of objects and space is not a feature peculiar to hungry somnambulants. A soccer player experiences the "game of soccer" in much the same way. The player runs, stops, twists, turns, and kicks, all in response to a changing environment where the moves are "instinctive" or "ingenuous" rather than calculated and predictable. Faced with A Huge Halfback With A Bad Attitude, one doesn't look down to check where one's feet are with respect to the ball. The feet know what to do. That part of the experience of playing the game is woven into the player. What distinguishes a 'good' from a 'bad' player at this point is whether the player can seamlessly adapt to the new situation. The 'good' player isn't 'good' because s/he is a walking calculator of the 4567 possible outcomes of this situation. To deliberately use what is after chapter 3 a loaded analogy, the good player can like a river flow around that obstacle. And once around "Bad Attitude", the good player can realign with the new configurations of the field.

The example of the "game" points out the new kind of "object" we need to be concerned with when we look for Complexity in human artifacts. The game is not a thing itself, it is not a set of rules and a few hominids and a playing field. It exists in the playing. The ball and other players are in some sense objects external to us, but we are involved in a give and take with these objects. The bounce of the ball is indeterminate. That halfback's next move is not predictable. Somehow, we respond to each. While our ability to play is based on our physiology and on our knowledge of the rules of the game, what actually happens on the playing field is a response to something 'bigger than us'. This something is the game itself. It is a non-physical entity that encompasses rules, things like balls and shin pads and players, and mental events like intentions and responses to events. The game emerges from the event of its playing. In that respect, it exhibits one crucial characteristic of a Complex system.

And the example of the "game" also reaffirms that the new paradigm presents us with different ways to "study" such entities. In the previous chapter, we looked at the different goals of understanding behind the pluralistic methodology of Harvey and Reed. Here we have a less academic, but perhaps more poignant example of what it means to understand a system. We must view from within as well as from without. If we restrict ourselves to watching the game as spectators, we miss the experience of the playing itself.

The "game" can be viewed as a set of events; we can model it as a set of data points and graphically represent it in some kind of Soccer Field Space. We can break the events into "plays" and use these rules for another contest. But the system that is the game *as played* is not accessible to coarse grained study. For the period in which the game is played, the game is something quite different from the "thing-pieces" and the "rule-pieces" and it is accessible, *qua* game, *only from the inner perspective of the player*. Its complexity is based on the lived experiences of the playing, not on any set of computations about the options each player can take at each moment at every location on the field. The spectators may *post facto* examine and model the game as a collection of events involving shin-pad bearing point masses traversing the field of play. But their "object of study" doesn't exist unless the game is played, and as a game, it must be played, not play-acted. The lived experience of playing, in which "the game" emerges, is prior to, and prerequisite to, any analytic modeling.[30]

STRUCTURES IN EXPERIENCE - 2: FROM GAME TO LANGUAGE

In *Shadows of the Mind*, Roger Penrose finds himself in an intriguing quandary. He firmly believes science must come to terms with consciousness. And he argues effectively that the purely physicalist interpretation of consciousness is inadequate. But this leaves him asserting that consciousness is accessible to science, just not science as we know it today. Penrose's main concern is to outline the "new physics" that might link brain with mind, biology with our sense of Being. But along the way he makes a statement about the phenomenal experience of understanding that is worth pursuing here. For Penrose, our awareness of an object is prior to, and prerequisite to, our understanding of it – whether that understanding is scientific or non-scientific (Penrose 1996, 38-39, 53). And it is a shared awareness that underlies communication. This shared awareness "allows us to have some kind of direct route to another person's experiences, so that one can 'know' what the other person must mean by a word like 'happiness', 'fighting', and 'tomorrow', even though explanations are likely to have been quite inadequate" (Penrose 1996, 53).

In this, Penrose approaches a position held by the phenomenologists Maurice Merleau-Ponty and Hans-Georg Gadamer, though I should point out that Penrose's agenda in *Shadows of the Mind* takes him in a different direction than the one we will pursue here.[31] For Merleau-Ponty, both our experience of the world around us and our communication with other people are 'cooperative' efforts. We reach for the refrigerator door handle and find it at 3 a.m. in the darkness because spatial relationships are defined by our incarnate experience. They are part of the "structure" of experience. We speak to another and our words have meaning because they are part of a structure – language – that is something we possess within us, and which possesses us. It is not that words "have" meanings; rather, meaning "is the lateral relation of one sign to another which makes each of them significant, so that meaning appears only at the intersection of and as it were in the interval between words". Meaning occurs when words are understood in "profile", seen against the backdrop of the context in which the words are encountered (Merleau-Ponty 1964b, 42).

The "direct route" of which Penrose speaks is actually mediated by language, but because the meaning of words and gestures is based both on incarnate experience and on the encompassing structure of language, the mediation is "transparent":

> [Any sign] "... expresses only by reference to a certain mental equipment, to a certain arrangement of our cultural implements, and as a whole [it is] like a blank form we have not yet filled out, or like the gestures of others, which intend and circumscribe an object of the world that I do not see." (Merleau-Ponty 1964c, 88)
>
> "It is through my body that I understand other people, just as it is through my body that I perceive 'things'. The meaning of a gesture thus 'understood' is not behind it, it is intermingled with the structure of the world outlined by the gesture, and which I take up on my own account." ... "The linguistic gesture ... delineates its own meaning." (Merleau-Ponty 1962, 186)

Gadamer's description of how "language" provides structure to the external world is similar to Merleau-Ponty's (though his main concern is how language creates and informs our historical understanding). "Language is the fundamental mode of operation of our being-in-the-world and the all-embracing form of the constitution of the world" (Gadamer 1977c, 3). Language structures the act of understanding itself, so it preconditions our understanding of particles and people, sociology and sociable conversation, mathematics and menus. The phenomenological position makes room for both emergence and stasis in language. It does not deny the accuracy of lexicons and grammars, or the validity of Structuralism and Speech Act Theory. These treat language as a thing. They project "thingness" onto something that is not object-like and in fact *is prior to the notion of "object-like" itself* (Ricoeur 1974 and 1976). In the terms used here, lexicons and Structuralist analyses focus on components, not on the system that precedes and comprises the components.

A language evolves over time. Entries are added to the lexicon, colloquialisms change the rules of grammar, people ingenuously discover new 'fits' between 'words' and 'things'. In its evolution over time, language is emergent. Since language also reveals the world to us, as it emerges and evolves so does 'the world' - where 'world' is not the collection of objects one can stub one's toes on, but how those entities are understood, valued, and used. Gadamer's word for this world is "tradition" (or sometimes "history"). If "understanding is language-bound" (Gadamer 1977c, 15), then

> "Every interpretation of the intelligible that helps others to understanding has the character of language. To that extent, the entire experience of the world is linguistically mediated, and the broadest concept of tradition is thus defined – one that includes what is not itself linguistic, but is capable of linguistic interpretation." (Gadamer 1977a, 99)

If this sounds a bit too mystical, it's probably because we seldom think of language as anything more than a tool. Yet this "system" of language is similar to the "systems" sociologists and historians study: a social structure establishes to some extent how the people behave, what they believe in, what they think and buy, eat and wear. If I work in a software development team in the financial sector, it is 'probable' that I will be wearing business attire at work rather than jeans. If I speak 20th century American English, it is unlikely that I will utter "thanks so very much" when I bash my pelvis against that bureau in my hurry to get to work some morning. It is not impossible for me to wear sandals to work, it is not logically necessary that I say [enter your favorite expletive here] upon impact with the bureau. What I do as a component of a social system, what I say when in possession of and when possessed by a linguistic system, is one of many possibilities. It is in the doing, and in the saying, that the systems are *real*ized, just as it is in the playing of the game that the game is realized.[32]

STRUCTURES IN EXPERIENCE - 3: HORIZONS

I pointed out in chapter 4 that when a new culture wanted to 'enhance' Stonehenge to suit their particular religio-astronomical requirements, they had to contend with the existing structures and the physical boundaries of the geographic horizon. The placement of the standing stones allowed some celestial events to be visible but they also blocked out others. Within these constraints, each culture used sight lines within Stonehenge to mark the location of the rise, or set, or standstill of various celestial objects. The structures of Stonehenge became tools for focusing attention on significant celestial events, for highlighting them against the backdrop of the constantly changing night sky. Year after year, the constellations changed with the seasons and the 'wandering stars' we now call planets traced their sometimes circuitous paths across the night. And year after year, as cultures rose and fell, Stonehenge drew people's attention out to the horizon where they could actually see order in the midst of this chaos, constancy in the midst of uncertainty.

If the phenomenological analyses of perception and imagination are correct, we too have fundamental structures that highlight "objects" in our fields of view. And we too experience such "objects" as objects in part because they are called to our attention from the field of view. As you read this book, your attention is directly focused on the lines you are reading. But the rest of the room has not disappeared. The chair you are sitting in is probably partly visible in your peripheral vision – when you attend to that vision. But it is also present to your consciousness – it is comfortable, that's why you can think of reading rather than having your attention focused on the pain in your back from an uncomfortable chair. If you look up from your book and stare out the window at a house, you will see "through" the window to just the house. And it is a house you see, not a patterned colored surface. The "house" is presented to your perception, even though you cannot see the other sides of the house or the insides of the house.[33]

Merleau-Ponty describes this constitutive aspect of the perceptual horizon this way:

> "... the inner horizon of an object cannot become an object without the surrounding objects' [sic] becoming a horizon.... The horizon, then, is what guarantees the integrity of the object.... The object-horizon structure, or the perspective, is no obstacle to me when I want to see the object: for just as it is the means whereby objects are distinguished from each other, it is also the means whereby they are disclosed." (Merleau-Ponty 1962, 68)

Complexity and the Ordering Principles of Experience 119

In an act of perception, some aspects are immediately given to awareness and some are not immediately given. Both are required, both make up the "object" we perceive. There is the internal visual horizon of the house we perceive, but there is another, cultural horizon that helps situate the house on a particular street, in a neighborhood with certain cultural norms and social behavior patterns. The 'spheres of influence' used analytically in chapter 4 have a phenomenological analog in the nested horizons within the act of perception. The field of vision includes objects bounded by visual horizons, but there are other horizons that contribute to each object's 'content'. The object is a house, not a surface; it is a dwelling, not a wood box; it is the house of a friend, not that of a mere neighbor. All these aspects of the object are present to us in our perception of it, and they reflect the influence of our experiential context. This context contains ever broadening 'horizons' or 'spheres of influence' or 'prejudices' that are affected by our memory, our social situation, and our culture. The context provides the "style" of thought:

> "Perception is here understood as a reference to a whole which can be grasped, in principle, only through certain of its parts or aspects. The perceived thing is not an ideal unity in the possession of the intellect, like a geometrical notion, for example; it is rather a totality open to a horizon of an indefinite number of perspectival views which blend with one another according to a given style, which defines the object in question." (Merleau-Ponty 1964d, 16)

THE STRUCTURES OPERATING TOGETHER – PROJECTION & IMAGINATION

Science, like any other human activity, is essentially interpretive, "hermeneutical" in the jargon of Gadamer's intellectual tradition. Seeing, recognizing, and understanding are projections of our past experience, of our accumulated knowledge, and of our expectations into the present 'field of view'.[34] Back in chapter 3 you encountered a figure that could be seen as a fork with two or three tines. By the time you had adjusted to seeing both images, you could switch back and forth between seeing the two-tine fork and the three-tine fork. The interpretive or projective aspect of perception and understanding is similar to that 'mental shift' between the images: from the data on the page you constructed two different images, you structured the ink stains into a figure that had meaning for you. But we know that "structure" isn't restricted to geometric structure. Both incarnate experience and language can provide structure to the current experience.

At the beginning of chapter 4 you had the chance to stand with John Milton's Satan and survey your fallen legions. It is unlikely that you have actually ever seen Hell, though as I pointed out you may well have been through it in some recent development effort. Some of Milton's words, such as "woe", "sorrow", or "torture", may have no immediate correlate in your personal experience; and in fact a word like "woe" may not even be in your 'usual' vocabulary. Yet you can still understand Milton's words and Satan's plight because you can project your own experiences into the situation. You can imagine Satan's visage as he looks across the burning plain of Hell at hundreds of his followers writhing in the flames. You can imagine him calling to his troops to rally themselves for another assault. You can imagine his hateful stare upwards as he shakes a defiant fist at the Empyrean from which he was expelled.

Whatever images you created in your imagination as you read the previous sentences certainly had their origins in your experience with people and things. Your image of the plains of Hell might have been inspired by your experience driving through Death Valley in 120 degree heat. Satan's fist-shaking gesture 'looks' like a gesture of defiance in your imagination because it has been created from your own gestures. But in the event of imagining Satan's gesture, all the constituent perceptions and memories and incarnate experiences are formed into something new. These incarnate experiences and your linguistic and gestural expressions do not exist in the imaginal view as discrete, identifiable entities. The image that comprises them emerges from your experience of self and your experience of the world around you. It is not reducible to those experiences *per se*. The image is more than a pastiche of your past.

But there's something else of interest in your imagined Satan besides his pedigree in your experience. That image isn't like images on a TV screen or the view you see when you look at a house across the street. It appears spontaneously and it is indivisible. When you look across the street at a house, the façade, the lawn, the fence, etc. all have a distinct individual presence to your gaze. But were you to "focus" on your imaginal Satan's sword, it is a new image, not a part of the original 'full body profile' you might have had. And because attending to one 'aspect' of some imaginal view effectively creates a new and distinct view, there's a significant corollary here: imaginal renditions are not subject to tests for veracity or verisimilitude. Your imagined Satan *is* the Fiend himself.

Edward Casey has explored the phenomenological aspects of imagination in some detail (Casey 1976). His description of one characteristic of spontaneous imagination, "self-generativity", should bring to mind Goerner's description of Complex systems as "self-organizing" and "self-maintaining":

> "To be self-generative may mean any or all of the following: not subject to external coercion or control, but instead bringing itself about (thus the 'self-' of 'self-generative' refers not to the imagining subject but to the imaginative act or presentation itself); arising from an apparent lack of cause, motive, or reason; appearing in a way that is unsolicited and unpremeditated by the imaginer and emerging without any express effort on his part; appearing in such a way as to surprise; operating by means of its own self-propelling forces, being wholly self-determinative in this respect; and generating itself all at once, *totum simul*, without any significant prolongation or sense of steady development." (Casey 1976, 71-2)

Casey also explores "controlled" imagination, where we consciously will ourselves to call up an image. In that case, there is an apparent "cause, motive, or reason" and there is "effort" and there is likely to be no "surprise". But these are attributes of the act of controlled imagining that sets it apart from the event of spontaneous imagining. For our purposes, I want to examine how some controlled imaginative acts are still structured by the image: that is, they are in some respects self-generating and self-maintaining.

If I ask you to imagine yourself getting up from your chair and walking across the room you are in right now, you can certainly engage in a controlled imaginal effort that takes your phantom self from one location to another. The "scene" in which this brief imaginal play takes place is defined and delimited by the room you are in. In a sense, you control the actor in the play but not the scenery. Similarly, when you imagined Satan for the first time, it was a spontaneous act "seeded" by the words you were reading. John Milton's words established some of Satan's character traits for you and set some of the details of the scene. The rest was 'up to you'. If I ask you to "focus" on Satan's sword, you will still be 'creating' a sword, ingenuously, from your collective experiences. Your act of imagining constitutes the object you imagine, but even though the act has a consciously intentional "cause" the object revealed in that act does not. The act of imagination I just asked you to perform was calculated and controlled; but in that act, the ingenuous ingenuity of imagination projected on your 'mind's screen' a sword through, and as it were in spite of, your act of will.

HISTORICITY AND COMPLEXITY IN COGNITIVE AND LINGUISTIC ARTIFACTS

If by now you have the impression that we live our lives in "windowless cubicles" and spend our days playing a virtual reality game where we are designers, developers, and end-users all rolled into one, that was certainly not my intent. The structures described above are not our personal possessions; we are their incarnations. The game exists in the playing, in the players; language is realized in utterance and in understanding utterances; in play and in discourse we are their emissaries. Projections and imagination instantiate individual experience in new forms, while the shared systems of language and culture ensure the pedigree, authenticity, and intelligibility of those forms.

It is still an interesting question how we come to understand that these structures are at work and how we come to understand they have in some way misled us. For Gadamer, we become aware of our "prejudices" only when something goes wrong or something calls attention to how we are behaving.

> "One of the fundamental structures of all speaking is that we are guided by preconceptions and anticipations in our talking in such a way that these continually remain hidden and that it takes a disruption in oneself of the intended meaning of what one is saying to become conscious of these prejudices as such." (Gadamer 1977b, 92)

What is true of speech is also true of reading, of perception, and of understanding in general. What William Derham saw as a window on the

Empyrean we see as the Orion Nebula; where Newton saw a classically bounded universe we today recognize Chaos and evolution. At the risk of over simplification, paradigm shifts are what happens when one set of preconceptions and their accompanying linguistic expressions are replaced by another set of conceptions and expressions.[35] The history of paradigms suggests that perception and understanding have an historical dimension: that is, what we perceive and what we understand evolve over time. In this historicity, in this evolution, lies the key to Complexity for a particular class of non-physical, software systems.[36]

Suppose for the sake of argument we say that a physical system is Complex when it evolves through some mathematically specifiable generative algorithm, perhaps periodically enduring Chaotic transitions. At any given point in the system's evolution, an observer will see only the system as it presents itself at that moment. If the generative algorithm is such that it is completely immanent in and derivable from the system at that time, one could in theory create a model for the system that was accurate throughout the life cycle of the system. It would not matter at what the point in time one developed the model.

But when developers create a product, they are participating in a unique event. I am referring here not to the product, but to the production process. When developers turn requirements into code they are engaged in an utterance. Like speech or thought, this act is a singularity, and it is irreversible. We can't experiment with it, we can't reproduce it to determine trends, and we can't lump it into some class of events to probabilistically determine its precise future characteristics. We can, of course, approach the product in precisely these ways. But historicity isn't an attribute of things; it is an aspect of understanding. As Prigogine puts it, "Becoming is the *sine qua non* of science, and indeed, of knowledge itself" (Prigogine 1997, 153).

If thinking could somehow be reduced to physical systems or to deterministic computational efforts by those systems, then one might be able to turn the singular event of thought (or utterance) into a predictable instance of a certain class of events. Such "physicalist" or "computational" or "reductionist" theories of mind and mentation enjoy much popularity, but they also have significant opponents, such as Roger Penrose and Steven Horst. To delve into this many-faceted debate would take us far from our planned course, but before we look into encoded utterances, let me place this discussion in the context of that debate at large.

I confess I am persuaded by Horst's arguments against some forms of the Computational Theory of Mind, in particular his analysis of the semiotics of statements about mental processes and states. On Horst's view, a mathematical-functionalist theory of mind or language might succeed in creating a model of mental or linguistic behavior. But its intelligibility as a *model* means it would have meaning only by implicit reference to the very intentions and mental states it purports to explain independent of intentions and mental events (Horst 1996, 163ff). While Horst may not have had the last word in the debate, I will assume here that his critique of physicalist theories frees us to examine cognitive, expressive artifacts using the experiential strategy, with the following caveat. We will not follow Horst's course – he is interested in justifying a limited version of the Computational Theory of Mind using the phenomenological experience of mind as, so to speak, a benchmark. We, on the other hand, will probe the structures of experience and look for the source of Complexity in that cognitive, expressive artifact we call software.

Penrose's approach is more speculative than Horst's, but in many ways it is more intriguing. Horst is content to allow computational theories more time to further develop their models even as he shows that such models will be inadequate no matter how well formed. Penrose instead argues that the computational model for consciousness is inadequate even on mathematical grounds and proposes instead that the key to understanding mind and mentation lies in better biophysics not in better chip fabrication. He believes we will need a new physics to understand mind, a physics that can better explain some of the currently problematic aspects of quantum theory especially with respect to the biophysics of the mind. For Penrose, mind occurs in the entanglement of deterministic quantum with random local factors at the neuro-biochemical level.

Allowing for some depreciation of Penrose's argument when exchanged into the terms we have been using here, it seems to me there are interesting 'structural' parallels between the deterministic physics in the microtubules and the indeterminacy of the environmental forces, and the self-generative nature of the imagination and the constraints it works within based on our experience and on what is imagined. Perhaps one day we will have a hierarchical model of thought that shows the 'broken symmetries' between neuro-physical and imaginal "events". In the meantime it is tempting to suggest that meaningful utterance (or understanding one) is an "event" in a sense similar to the fecund and perplexing quantum "event".[37] The utterance is the realization of projection and horizon. One-time-only and irreducibly, it manifests our past and our present, it collapses all the possibilities of who we are and what we might express into a single actual utterance.

In the interplay between projection and horizon, both of which are themselves evolving through our lived temporal experience, lies the key to the kind of Complexity we encounter in software systems. Language and thought are structural "ordering principles" rather than causal factors. As the interrelationships within the language are revealed in and by our experiences and utterances, that structure 'grows'. As our perceptual and cognitive abilities are revealed to us in and by our experiences and utterances, that structure 'grows'. The result is that the ordering principles of thought and language are intertwined with the experience of the being whose world they help form. Incarnate experience, language, and thought are, in the jargon of Chaos, "coeffecting variables". This coefficiency, this interdependence between structures that is revealed only in time and through time, makes understanding (thought, utterance, imagination, etc.) emergent. Complexity, at least where cognition and expression are the primary constituents of the "system", derives from the nature of thought and language itself.[38]

From Utterance to Code

The thoughts and the "utterances" of software development are constrained in such a way that the potential for Complexity is less in coded utterances than in 'living' languages, but the structures of experience nonetheless operate in software development and potentially contribute to software Complexity. In anticipation of the next chapter, let's explore this potential using an illustration from that aspect of the development effort near and dear to the hearts of software professionals – generating and coding to requirements.

Building a product incrementally with periodic customer reviews of the 'prototype' and its progeny allows the development team to verify at defined stages in the project that the product it is building is the one the customer wants. Using the fundamental structures of experience described above, we can explain why this "approach" can be successful.

Anyone who has negotiated customer requirements knows that much of the effort in requirements management occurs in determining what the requirements really are. The usual complaint goes something like this: "What the customer said did not accurately reflect what the customer actually wanted." That's often a surprising statement for those outside of the software industry, and the surprise sometimes is voiced as the objection "Who are we to judge what the customer 'really' wanted?" Of course, neither way of looking at the situation is correct. The customer wasn't lying, they told the development team exactly what, in their mind, they needed. The development team, as it analyzed the requests, determined that the exigencies of platform, time, good programming practices, etc. forced a "refinement" of those requests. The horizon in which the customer expressed his or her requirements was different from the horizon of the developers. The "preconceptions and anticipations" on either side of the conference table were different.

We've seen these horizons at work in chapter 4, such as when the end user's appreciation for the product's performance is affected by the work place environment and by the user's previous experience with the product. In this case, the user's incarnate experience can be a significant factor in how the product is perceived. A harried office environment highlights the slightest delay in a product's response; compared to a previous version, the input fields are inconveniently placed for the tasks this user needs to complete. The pressure the user feels is transferred to demands on the product. Such 'requirements' are probably not known as the customers sit with the development team around the conference table. But as we have seen they can be a significant factor in the system of the software as executed, and they can contribute to the unpredictability of how the system will behave and how the product will be received.

Of course one of the benefits of prototyping is that it gives the users the chance to 'see and feel' the product they originally only imagined in the requirements phase. The requirements expressed at the conference table reflected what the customers then envisioned as "the product" and as "their needs". Their 'visions' of the product and their needs are certainly acts of controlled imagination rather than spontaneous imagination. One would expect that the screens and fields and performance attributes they envisioned would be guided by specific past experiences; the "scenery" for their mental images is less random, more accessible to others. But just as we are not aware of our prejudices until our expressions and understanding fail (Gadamer 1977b, 92), so is it necessary for customers to actually 'test drive' the product they specified to find out whether the requirements were "accurate". If the product fails to perform as expected, it may be because the requirements did not express to the development team all that the customers intended, or it may be because the customers did not envision all that they ultimately wanted in the product.

Suppose the requirements accurately expressed the customers' intentions. Can our analysis of language and understanding help us explain why the developers might still "get it wrong" in this case? Listening to or reading the requirements, or examining the design specification, the developers begin to form a mental image of 'how the code will work'. This may include snippets of routines they have already written which they can reuse. It may include some sort of imagined flow diagram. If protocols are involved it may even include imaginary 'conversations' between nodes. The developers then express their understanding of the requirements and design in linguistic expressions we commonly call "code". These expressions are not precisely the same as spoken or written utterances in the native language of the developers, but they retain some significant features of these more natural expressions.

First, they are intentional expressions. Lines of code embody directives, they have referents: the developer refers to and manipulates fields and values, the developer 'signals' some other process and passes to it information that will prompt that process to perform some act. Second, these expressions are constituted by the developer's experience and by the capabilities of the programming language used. But as with the customer "view" of the product, the developer's "view" may be deficient either in vision or in expression. As the developers type in their directives and references, they are speaking obliquely on behalf of their view of how the program should flow and how the product should function. This view is both cognitive and imaginal, and when this view is realized in the directed use of language there's a possible tension between the self-generative nature of imagination and the constraints of goal directed utterance. No one sets out to create a bug. Developers believe that what they are encoding will have the intended effect. When I say, "Please pass the salt" I don't expect to be handed the pepper mill.

The problem is, programs aren't one line long. They are complicated discourse, module to module, process to process, developer to developer. Anyone who has received a D on what they thought was a brilliant term paper, or who has had the pleasure of bringing a major customer to a standstill with code that passed a rigorous peer review, knows how imprecise communication in words or in code can be. As the product is built up, the complexity of its linguistic content increases. Ideally, the coded functions 'flow' to their intended (required) completion, just as a term paper ideally leads the reader through its arguments to its startling and brilliant conclusion.

But unlike the term paper where there is one author, one mind directing the outcome, in the program there are many interlocutors, as modules interface with other modules, processes signal processes, etc. As we shall soon see, some of the characteristics of potentially Chaotic software systems also occur in Complex software systems. When program logic is in fact dialogic, and when the program resembles a round table discussion rather than a monologue, the structures of experience operative in the developers set the stage for emergent behavior in the executing software system.

CHAPTER 7:
COMPLEX SOFTWARE SYSTEMS

> That's not a bug, it's a feature.
> *Anonymous*

Anyone who has spent any time in software development or product maintenance has heard that line. Some of us have probably even used it ourselves. Sometimes it is meant as a joke – although the humor may be lost on the irate customer on the other end of the technical support phone line. Sometimes it's an excuse, as developers try to present a serious functional oversight as something intentional, deliberate, and pre-planned. But sometimes it is a serious, and revealing, statement about what we know about a software product before its release, and what we know about that product after its release.

In typical industry parlance a bug, no matter how it is classified or when it is discovered, is generally acknowledged to be a mistake. But "feature", used in the serious sense, seems to occupy a different order in the phylum of software defects. Test engineers tell developers that their code has a bug; they don't rush down the hall to tell them that their code has a "feature". Developers fix bugs with determination, perhaps even with repentance for a mistake they made. On the other hand, they approach "features" more cautiously. That kind of defect *might* be an oversight, a "gotcha" or some other variety of larval liability in their code. But it also *might* be an implication or corollary of a design decision that could not have been envisioned during the design phase. In the latter case there's a different degree of culpability. Writing a bug is cause to slap one's own wrist; discovering a feature is cause to wonder just how it got into the code in the first place, and what to do about it now that it is there.

One could certainly take the position that such "features" aren't features at all. They certainly aren't like the print menu feature that was designed, implemented, and verified according to plan. From the customer's perspective, this so-called "feature" is a bug, plain and simple. But it's not the customer's perspective that is intriguing here. That developers with (more or less) straight faces would call a defect a "feature" reveals more about software behavior than it does about the behavior of developers under fire for their mistakes.

A feature is a product function or characteristic whose genesis can be traced to someone's intentions. The customer wanted it this way; the developer coded it this way. The proper operation of the product is a natural outgrowth of deliberate, focused, productive actions during the development process. The behavior is "correct" in that it meets specifications. The behavior is "justified" in the sense that developers can retrospectively grasp that there is a relation between the observed behavior and the product they created. The use of "feature" to refer to a defect suggests that there are some kinds of product deficiency that are also "justified" in retrospection. They may still be defects, but they are not mistakes. They still require remediation, but in a significant sense the remedy includes new *experience*, a new appreciation and understanding of "how the code *really* works."

This kind of "justification" is often seen in historical analyses. Few would deny the contribution of Issac Newton to science. Fewer still would claim that his view of active and passive influences in a universe directed by God's Will was a "bug" in his theory, a mistake for which he can be blamed. Historians are more likely to trace such features of the Newtonian world view to the social and intellectual context in which he lived – to religion, to Aristotelian theories of causation, to the alchemical tradition that lingered on even in Newton's time (Heimann 1994, Guerlac 1994). One might be disappointed that the author of the *Critique of Pure Reason* could do no better, but when Immanuel Kant located the center of the universe at the star Sirius, he was merely following a centuries old "logic" that linked magnitude (in a variety of senses) with physical or theological importance.

Such justification isn't rationalizing or excusing, nor does it pretend to be deterministically causal in some way. Newton's intellectual environment didn't force him to make matter a passive principle, but he did so because it "made sense" in his theory. Kant's *Cosmogony* shows at times a decidedly "modern" method of analysis in spite of the "legacy" thinking that made Kant look to Sirius as the center of the Milky Way. When we place a decision or action or utterance in a context, we acknowledge relationships and dependencies between the ideas and information available to the agent and the agent's behavior. To use phrasing from the previous chapter, we try to reconstruct the horizon in which the decision or action or utterance took place. It is not so very different from how we see ourselves in retrospect.

When I review my own behaviors yesterday, I see myself influenced by what I saw, heard, did, desired, believed, and feared. I don't see myself as the instantiation of some physical or social laws, a bipedal mass unable to escape the gravitational pull of some preordained future. I move intentionally, but within constraints set by my physiological and physical environment. I express myself, but convey only what the language I use will allow me to convey. I *make* rather than have decisions, fashioning them from scraps of previous experiences, concepts, and imagined outcomes into a whole that is different from its individual parts. My horizon - the combined effects of physiology, physics, language, experience, knowledge, desires - delimits my behavior but does not cause it. The interdependency between horizon and what transpires within it is not deterministic, it is enabling. It is like the relationship between the player and the game: the game (its rules and the act of playing it) constrains but simultaneously enables *playful* behavior.

Am I suggesting that a development team is being playful when they inadvertently create a product characteristic that one day they will call a "feature"? In a way, yes. We all make "bad" plays in a game, though we aren't aware of the deficiency of our actions at the time. It's only through what happens after our play that we realize it was a "bad" play. Their minds filled with project deadlines, product specifications, and their own coding styles, the developers respond to those influences like halfbacks to the advance of the opposing wings. Like Kant, they are possessed by certain habits of thinking they may not be aware of. Like Newton, they solve problems and resolve inconsistencies within the framework or structure or horizon of their development experience and product knowledge. Time and experience may show a developer the error of his ways; but at the instant the developer hits the Enter key and deposits the "feature" into the source code, that developer is responding ingenuously and correctly to the influences of the moment.

When, in retrospect, that same developer calls some unanticipated product behavior a "feature", the developer is acknowledging that in this case at least software behavior is, from an experiential point of view, non-deterministically emergent. That behavior is not what the development team expected, but with 20-20 hindsight, it "makes sense" that the software would behave that way. The potential Complexity of software lies not in the difference between the expected and actual performance of the product, but in the reasons why that difference exists. Precisely because code generation is an intellectual, linguistic activity, it is subject to the influence of Complex structures in experience and understanding.

THE FOUNDATIONS OF COMPLEXITY IN SOFTWARE

The notion of a "feature" described above may provide a useful introduction to various aspects of Complexity in software, but it doesn't explain what software Complexity is or how it comes to be. For that, we must understand what sorts of emergence and Complexity are possible in software systems, how the structures of experience identified in the previous chapter contribute to Complexity, and what role Chaos may play in the Complex system.

The notion of emergence in software is especially complicated because it changes according to the strategy used, the kind of product, and the anticipated behavior of the product. One case of software Complexity, that created by artificial intelligence programs, can be analyzed according to a coarse grained strategy and is likely to be deterministic as well as emergent. The anticipated behavior of the program includes its ability to process data from its environment and to "learn" to respond in various ways. Structurally and behaviorally, such a program is algorithmically Complex: from a basic set of intended behaviors (coded algorithms) the program self-generates other algorithms and maintains its stability and/or control inspite of environmental changes.

At the risk of irritating some researchers in whose view such Complex behavior is the only legitimate kind, I want to investigate another arena for emergence and Complexity. This kind of Complexity has its roots in time, in the structures of experience, and in the nature of linguistic utterance, rather than in self-generating algorithms. In that, it also presents a significant challenge both to the software industry and to the new science of Complexity when that effort is modeled on the physical sciences. One can show that this 'special' kind of emergence is compatible with the kinds described in the physical sciences.[39] One can even find evidence that the hard sciences are approaching a less analytic, more holistic position with respect to systems, knowledge, and evolution (Robertson 1995). Interesting as those discussions might be, they would take us far afield from our concerns here, the genesis and ordering principles of Complex software systems.

My first task is to defend the notion that the emergent behavior of software systems can be different from the emergence typical in physical systems. In the sense normally used, "emergent" behavior includes a system's ability to organize and modify itself, usually (in the case of physical systems) in compliance with some set of laws. But the coarse grained, subject-object perspective ignores the evolution of understanding, the emergence of *systemic* views from what was a component view. Where the system being studied is not object-like, but instead is an artifact of intention, *the evolution of understanding may be the defining characteristic of the Complexity of the system.*

CHAOS, COMPLEXITY, AND TIME

In a way, "Chaos" is more a matter of situation than of scientific method, more a result of how we see things rather than how they "really" are. The universe was congenially Newtonian - for a while. River rapids weren't Chaotic - until recently. As knowledge increases, as our ability to understand relationships and dependencies evolves, the things in the world around us, the laws that govern them, and our awareness of ourselves also change. Chaotic systems in 1700 were no more or less predictable than they are in 1998; the difference may lie in the fact that in 1700 we may not have encountered the *system* we now see in 1998, only some of its parts. Or the difference may lie in the fact that we expected, one might say insisted, that the world behave linearly and predictably, and so we dispensed with Chaotic systems as anomalies. When we could not comprehend the structure of Chaotic systems, when we could not break our habits of linear and deterministic thinking, the world *was* linear and deterministic. Our cultural, historical, and scientific situation left us ill prepared to perceive, let alone understand, nonlinear, nondeterministic systems.

But once the paradigm shift had begun, the world changed with it. If one gives the same validity to applications inside the social sciences as to those inside the physical sciences, the "history of Chaos" shows how quickly Chaos and Complexity have evolved from the study of indecorously behaving physical systems to an ideology that finds Chaos and Complexity in tree branches and bronchial branching, in mentation and mental illness, in the family room and in the corporate board room, and everywhere in between. Our knowledge of Chaotic systems in general (not just their special incarnations in physical objects) *has itself evolved*, from weather to worldview, in just a few decades.

The "history of Chaos" has an analogue in the historicity of our understanding of systems. A system may not be predictable, period. But it also may not be predictable because we don't know how it evolves - we haven't yet seen it evolve *as a system*, or we have no idea how its components evolve and affect the system. Its behavior may not appear to be linear because we have not adjusted to thinking about it on its own terms. It is dynamical as much because our perspicacity is limited as it is because the system itself can structurally generate new behavior. But once we have models to apply to systems, once we have grown accustomed to thinking about Chaotic systems as structured behavior, they become intelligible, maybe even predictable. Chaos becomes "deterministic" Chaos. Chaos exhibits "order" again. It is one of the salient features of the human mind that, over time, it can assimilate the novel, the challenging, the disconcerting, into categories it is more comfortable with.

I am suggesting that when we try to describe Chaos or Complexity, we should not forget that time and the revelation of systemic behavior in time play a significant role. Certainly the mathematically precise Chaotic systems in physics are more "lawful" than their sociological or psychological brethren. But Chaotic systems (even physical ones) reveal themselves in time, they are understood in retrospect as Chaotic systems. When an apparently linear system becomes recognizable as a Chaotic system, it may not be that it has somehow changed. It may be that we have seen it differently. It's not that we missed the possibility that the system could be Chaotic, it isn't that we missed the evidence. Rather, we didn't know that the system could be nonlinear, there were no grounds for assuming it could be. We couldn't see this "feature" of the system until it manifested itself.

If we look at serious, systemic failures in software products, e.g., the network wide system "meltdown", we can see this temporal limitation at work. When the failure manifests itself we understand it, not before. In retrospection we may come to understand how the failure was latent in the design and the code. But in this case we are actually rewriting history in light of the present in much the same way that memories, with their dual 'partiality' described in chapter 4, present to us who we are in the guise of who we were. We have assumed that the knowledge we have today, after the meltdown, is the same knowledge available to us when the design was approved or the code checked in. I suggest that that may not be a reasonable assumption. Didn't we inspect the design document? Didn't we inspect and test that code? Were we blind, stupid, and incompetent? Or were we designing and coding more than we knew at the time?

I am sure that those of more mechanistic persuasions will reply that the developers of the software that melted down simply failed in their duties. After all, if they had just applied [enter your favorite silver bullet here] they would have avoided catastrophe. But after chapter 2, you should recognize that such objections actually place the burden of proof on those leveling the objection: they must prove that there is a linear causal relationship between the silver bullet and zero failure product or the objection carries no weight. And after the discussions of spheres of influence, systems, and the strategies and their limitations in chapter 4, we know that the accuracy and precision of silver bullets is extremely difficult to prove, even without the influence of Chaos or Complexity.

But after chapter 5 we know that there *are* certain characteristics of software products that can support Chaotic behavior in the product. And we know that some of these characteristics are inextricably tied to intentions and expectations, which places them out of reach of any tools in the development manager's toolbox. We learned in the previous chapter that there are fundamental structures in our experience from which our thought and actions emerge. We should therefore expect to find the "special" form of software Complexity we are seeking in the realm of understanding and utterance.

CHAOS, COMPLEXITY, AND UTTERANCE

In chapter 5, we examined one software system - the execution, in a specific environment, of encoded components (statements, directives, and protocols) for a specific purpose with anticipated operational outcomes. Within that system, we defined the 'equilibrium state' of the product as the product operating in compliance with macrolevel and microlevel expectations. By extension, we identified far-from-equilibrium states as cases where the actual performance of the product differs from the expected performance or where the actual code paths executed veer from the expected code paths.

But a far-from-equilibrium state, so defined, is not enough to make a system's behavior Chaotic. We needed definitions of nonlinear and dynamical behavior to differentiate Chaotic from improper program behavior. Although the goal of chapter 5 was to transpose the concepts of Chaos from the hard sciences into the realm of software, we found that we could not do so without paying some attention to the intentional side of experience, to expectations and assumptions about how the product will work. As a result, nonlinear behavior was defined as program behavior that deviates from its intended 'logic' or 'flow', that is, when some microlevel expectations about behavior have been violated. Dynamical behavior ensues when nonlinear program behavior spawns systemic behavior that violates the purpose of the program and/or changes the anticipated outcomes of its operations. I also suggested that the attributes of hardware affinity, dialogic functions, and operational complexity were the product characteristics most likely to contribute to Chaotic behavior. And I attempted to show how, given a specific instance of one of these attributes and given some triggering series of events, the system's behavior can become Chaotic because of the way these attributes structure systemic behavior.

As it so often does in other systems, Chaos prepares the way for Complexity. If we now combine what we know about the ordering principles in experience with what we know about Chaos in software, we should be able to describe in more detail how intentionality contributes to Chaotic and to emergent Complex behavior. In the previous chapter, we learned that linguistic utterances issue forth from our own internal horizons: within the constraints of language, experience and knowledge, they are the realizations of our intentions and projections. Normal conversation occurs under fewer constraints than does coding in software, but both are linguistic expressions and both are subject to the structures of our lived experience, the context in which we speak, and the structure of the language used.

When I ask you to pass the salt, I expect you to understand me and I expect you to pass me the salt shaker. When I encode a print directive, I expect the program to print the specified file at the appropriate moment in program execution. The additional level of indirection in the software statement does not change the fact that the encoded utterance has a purpose. It expresses an intention on the programmer's part to cause something to happen, and its location in the code flow signifies the programmer's anticipation that the intended event will occur under certain conditions.[40]

Whether oral or encoded, linguistic expressions are imprecise. As my conversation emerges out of my horizon in the event of uttering, and as it is absorbed into yours in your act of understanding my expressions, my intended and anticipated effect ceases to be mine and may not even occur. Should you jump up from your chair, back up towards the wall, and fire the salt shaker at me with the aplomb of a professional football quarterback, I will conclude that you misunderstood me. But you did respond appropriately, given your understanding of my directive to "pass" the salt. Understanding, in conversation at least, is a shared responsibility and carries shared liabilities.

THE COMPLEXITY OF ENCODED UTTERANCES

With encoded statements there is no human interlocutor, but the software execution functions as an "agent" whose computational and environmental characteristics approximate that of the people we converse with. The flow of executed statements sets up its own context for each ensuing statement execution. That context may be simple or it may be complicated: it may be a 'hello world!' program or a multi-threaded effort to partition detailed calculations across multiple processors. The current hardware and operating system environments function both as supporting and as limiting factors to the effectiveness of the executing statement. They help complete the programmer's intention implicit in the encoded statement, or they may somehow alter its execution and thereby stifle the intention. It is hard to allocate memory or store data when there is no memory available or the physical bus to the storage device is down. And for reasons already discussed in chapter 5, where there is dialogic activity between multiple processes on different nodes the potential for aberrant behavior is increased.

In chapter 4 we saw that when a developer sits down to turn the product and design requirements into executable code, she or he faces far more than the keyboard. There are the pressures and constraints of the project. There are supporting or annoying factors about the work environment. The developer's lifestyle, home life, and physiology also come into play as the directives and statements are encoded. To this list of 'initial conditions' that contribute to the creation of the code we now can add the phenomenological context in which the encoded utterance takes place. Thinking about how to get the program to do what the developer wants it to do, the developer imaginatively projects into the future, entering into a peculiar sort of dialogue with the silent entity of some execution event, at some future run-time. Individual coded statements may appear to be imbued only with the developer's understanding of the requirements and design, devoid of significant imaginative or projective context. But just as "Pass the salt" is cogent only in the context of a conversation over food, so are these individual encoded utterances meaningful or significant only within the context of what the code or function or module is intended to do.

We don't have to stay in the rarefied atmosphere of philosophy to find support for this hypothesis – we only have to listen to developers. Eavesdrop on developers talking about their code and you won't hear them speak the jargon of the processor logic. They don't talk of chips and gates and signals. Nor do developers speak the unmediated diction of the code itself (e.g., "paren paren const Label ampersand paren Label paren paren equals equals True"). Instead, they'll narrate their way through a task. "We've just received a status from a remote thread on this socket, and now we downcount the outstanding operations counter and store the pointer to the data buffer." This narrative "we" suggests that the developer in some way identifies with the *execution* – not with the code but rather with the intended operations, not with the encoded directives but rather with what the developer anticipates will happen at run-time.[41] This identification is a characteristic, one might say residue of, the event of code creation.

THE IMAGINAL AND THE REALIZED DIALOGUES

While I don't claim that all developers, in every case, engage in an imaginal dialogue taking place in some future execution event, when this does occur it provides a basis for emergent behavior. The developer creates directives and structures their relative execution in an act of controlled imagination. Such imaginings lack the pictorial quality of your image of Satan on the plains of Hell. They also lack the topographical constraints of your imagined trek across your room. But like those imaginative acts, the imaginal content is non-verifiable and indivisible. It is ingenuous and self-generating. The behavior of the code, its intended meaning, is a result of both imaginal projection and deliberate utterance: the ingenuously felt reality and certainty of the imaginal projection blinds us to any other possibilities in our utterance but the one we intend.

But if the utterance is deliberate, its effect may not be deterministic. Whether "Pass the salt" or "Pass the pointer", we express our intentions and anticipate our expected results. We are not aware of other possible results of our utterance until they happen; we are not aware of features until they manifest themselves in the execution event. The dialogue imagined by the developer won't be realized, the intentions behind the coded utterances won't be fulfilled (or thwarted), until the actual execution. At that moment, the developer's experiences, understanding, and coded utterances receive their actual expression and become subject to all the indeterminate, unpredictable affects of the run-time context in the execution event. This execution event is a multi-tiered dialogue – a dialogue between one line of code and another, a dialogue between one process and another, a dialogue between directive and anticipated hardware/software contexts, even a dialogue between developers represented by proxy in the code they have written that interacts at run time.

Software behavior can be both emergent and Complex to the extent that it is structured (enabled, constrained) by the code and by the intentions of the developer who may not foresee (imagine) all the possible outcomes of the directive. Put differently, the intent and the language in which it is expressed, and the exigencies and peculiarities of the actual run-time environment, coeffect and may create a new dialogue, a new behavior, that reveals unanticipated possibilities latent in the developer's act of utterance. To modify what was said in chapter 6, Complexity in this special case of software execution derives from the nature of thought and language in the contexts of development and execution. Complexity may be nudged into revealing itself by physical (run-time) factors; it may even require evident failures to be noticed *as* Complexity. But it is nonetheless the over-arching system in which the encoded utterance is merely one initial condition, the intended effect merely one possible trajectory, and the developer merely one player in the game.

ORDERING PRINCIPLES IN SOFTWARE COMPLEXITY

We know from chapter 5 that many 'simple' software programs are uninteresting when it comes to Chaos. That may be less true for Complexity, since Chaos is partly dependent on the interrelationship between execution paths and Complexity is partly dependent upon the developer's context and understanding. But for the sake of illustration, let's focus on a complicated, dialogic, multi-threaded, distributed application. The architecture for our mock application is divided into "nodes", so called because each of the segments executes on separate processors on separate desktops. The user interface is a thin client, through which users can request service from a dispatch node. The dispatch node performs operation segmentation and dispatches the segmented sub-requests to several server nodes that perform the actual 'work' of the application. Internode communication is based on an embedded communication protocol and a base class of methods which all objects inherit when instantiated.

By using a common off-the-shelf communication protocol and by limiting object exposure through private function-oriented methods, the design tries to simplify internode communication and to isolate objects from one another. It would appear that coding any given object would be fairly autonomic: given a coding convention, some experience in kernel-level C++, and a reasonably detailed implementation specification, the developer could simply "translate" into C++ the "requirements" of the design. I suggest however that this appearance is misleading in anything but the ideal case of development.

In the ideal case, the design is exhaustive in its coverage of availability and reliability and the developer will translate such requirements along with the 'success path' requirements. Part of the developer's context will already include directives, couched as specification statements, to 'say' this or that upon receipt of an exception or upon timing out another node. Additionally, in the ideal case the implementation statements are themselves couched in some formalistic metalanguage with a very close correspondence between the metalinguistic statement and the resulting code statement. With little left to the imagination (figuratively as well as phenomenologically), the developer will be performing transcription rather than uttering encoded statements. But if the developer is working from a linguistic or pictorial design where there is no close correlation between design and coded statements, the event of coding moves away from transcription towards utterance as the need for imagination and projection increases.

Watch a team of junior developers working on such a product without the ideal support of exhaustive specifications and closely coupled requirements and code (or remember your own 'learning curve' in such a project). You will probably find that unless the developers are extremely astute and more than slightly paranoid, error handling will improve incrementally as the product matures in the development and test cycle. The developer probably works through the main functions first. In order to encode some basic exception handling he imagines 'what could happen' in the code's execution environment and codes defensively for those envisioned events. Ask him what he's coding and he might answer "If we try to allocate a resource and fail, we'll throw an exception. If a server node throws, we'll just pass it back to our client." He is projecting himself into an envisioned or imagined 'conversation' and his response is encoded in the statements he eventually prints off for an inspection.

And in that inspection, someone points out that "You have to roll back your transaction to a known state for a safe restart when the client re-issues the request." This very reasonable advice is, however, still operating in the developer's original context. What has not yet been addressed are issues of data integrity on the server's side, timing dependencies across the nodes, and whether the client can handle the exception and reissue the request. The boundaries of the horizon in which the inspection takes place, the limitations on the dialogue in the conference room, are influenced by the encoded utterances on the print outs everyone holds in their hands. Not determined, merely influenced: someone experienced with distributed multi-threaded products would no doubt point out these and several other deficiencies in the current code. But that in fact serves to reinforce the notion that we encode from, or decode based on, our experience and understanding of the context of code execution.

Suppose the developer had skipped the inspection and handed the code over to the Quality Assurance team to test. Suppose that testing demonstrated that in error conditions the code behaved as intended but the client could not restart operations without incurring some form of data corruption. This is certainly unanticipated behavior: the developer envisioned passing the exception back, not corrupting data. But is this unanticipated behavior Complex – is it emergent? In one sense, it is emergent: it appears, unenvisioned, unanticipated (not to mention unwelcome) from an utterance that was intended to do something else (merely pass the exception). But one could argue that in this case "emergence" is little more than a lack of foresight. This case lacks the innate structure required of Complexity *in software*. The structures of personal experience – horizons, projection, language – are clearly operating here, but one could still claim there is nothing intrinsic to the software that caused product or system behavior that is somehow self-maintaining and self-generating.

We earlier presented far-from-equilibrium behavior as necessary but not sufficient for Chaos: to make behavior Chaotic, oscillations and turbulence had to involve more than one code thread, they had to involve the 'system' as a whole. Just because some module behaved otherwise than expected, we couldn't legitimately call that behavior Chaotic. Similarly, we can't take non-linear, far-from-equilibrium, or far-from-expectation behavior as the sole condition for Complexity. We can certainly expect software Complexity to depend upon personal horizons and imaginative projections and the encoded utterances they produce, in the same way that Chaotic behavior depends on non-linear, far-from-specification behavior. Every society has its deviants; some are criminals and some are visionaries and the difference between the two lies primarily in the ordering principles of the society itself.

So for there to be Complexity in software we must find "ordering principles" that are manifested in the execution event. In fact, the preceding analyses have given us most of the information we need. Combining the structures of experience, the special kind of utterance developers perform, and the exigencies of the execution event, we can define Complexity in software as the combined effects of:

7.1) The freedom and the constraints intrinsic to goal-directed, encoded utterances, and

7.2) The execution of interdependent encoded statements when those statements were generated as utterances and not as transcriptions.

Let's unpack this hypothesis in reverse, first distinguishing transcriptions from utterances, then giving an example of interdependent execution, and ending with an examination of freedom and constraints.

If the developers merely transpose into code exhaustive functional requirements, couched in some protocode requirements language, and if the requirements address all possible environmental contexts and prescribe all possible behavior, then the behavior of the software product will likely not be emergent (unless of course the requirements are to create an artificial intelligence that learns). The developers are transcribers, not speakers. They need not imagine how their module would/should respond to run time events; they need not identify with, or project themselves into, the intended behavior they then encode. When developers create a product lock step from such detailed specifications, there's little play between what is intended in the encoded statements and what will happen. There's nothing left for the software to do 'on its own' as it were. But where these ideal conditions are not met, where development is projection and utterance as much as production from specs, there is room for emergent, Complex behavior to the extent that the behavior of the software system was not constrained by the encoded utterances.

It is the systemic behavior that is important here. Nonlinear or unanticipated behavior that results in immediate program termination (where immediate is taken in the sense of either a chronological relation or a 'logical' relation in a control flow) isn't Complex anymore than it is Chaotic. It is bounded, terminal; the system doesn't evolve or adapt. Nonlinear, oscillatory program behavior with systemic effects is Chaotic. But if the turbulence doesn't result in a new stable systemic state, that system never becomes Complex. The system has to recover some form of stability. Multiple threads (or nodes or modules or code paths) must in combination produce a system state that is not immediately fatal and is not oscillatory. (And we must keep in mind that these threads or nodes or paths are enacting the discourse of the developers who 'wrote the script', even though at execution time they are acting independently of the developers.)

By way of illustration, let's enhance the design of the distributed application described above. To the basic functions already mentioned, let's add node recovery mechanisms. Let us suppose that if the dispatch node should fail, a new dispatcher will be selected from the surviving server nodes. A shared database is used to store current node membership and node status. The initial design calls for a time-out mechanism between nodes, so that if a dispatcher times out a server, or vice versa, that node can be removed from the membership database. If a server fails, the dispatcher simply restarts that server's outstanding operations on another viable server. If a server times out a dispatcher, that server node must force a reconfiguration of the distributed system, which includes the selection of a new dispatcher and notification to the thin clients.

This new design, simple as it may appear, gives the systems "decision making" capability. Algorithms in the code now determine such matters as time outs, node membership, and membership reconfiguration. There are two levels of dialogic activity here. At the lowest level are the protocols used between survivors: the keep alive mechanisms, the messages used to signal a change to the membership data, etc. At a higher level, there are the decisions made by the membership management code that are communicated through the lower levels. When one node signals a membership change, it has executed a code path that has "decided" that some other node is no longer accessible, and it signals its "intention" to remove that node from membership by changing the shared membership list.

But just as the intent behind "Please pass the salt" can result in unanticipated results, so can such dialogic directives. Suppose there are intermittent network failures that cause some nodes to time out some but not all other nodes. Server node A times out dispatcher node B, but server nodes C and D can still communicate with dispatcher node B. Node A requests a change of dispatcher. How will the software on the nodes behave? Obviously, that will depend on how well the development team thought through the various scenarios, but it will also depend on what they *didn't* envision. For example:

- If nodes C and D reject node A's request, what happens to node A?
    - Do nodes C and D wait for the dispatcher to remove node A?
        - Suppose the dispatcher can still communicate with node A, so nothing triggers the dispatcher to remove A?

- Suppose dispatcher node B times out server node C at the same time server node A times out the dispatcher?
  - Suppose both node A and node B initiate a membership change at the same time?
    - If the development team did not envision the possibility of simultaneous accesses of the membership list, they may not have protected the membership list from write collisions.
    - Even if they did, there are still "windows of opportunity" for disaster. Dispatcher node B changes the membership file before server node A does; but as nodes C and D respond to the dispatcher's request for a reconfiguration, they read the membership list modified by node A, not node B.

"Features", in the problematic sense cited at the beginning of this chapter, arise from failing to consider these and a myriad other possibilities in the executing system. Except in the case of naively optimistic designs, most development teams will have equipped their application's recovery logic to handle some subset of the possibilities. For some set of run-time eventualities, they will have successfully constrained the system's behavior with their encoded utterances: they ask for the salt, not the pepper, they want to remove a node, not crash the local system, and so forth. But when the execution event includes a multi-tiered dialogue and/or an unforeseen set of events, the constraints of goal-directed encoded utterances are broken and the executing system is free to evolve into unanticipated states. If the individual nodes in this example enter some internal loop states, connecting and disconnecting other nodes, the macrolevel behavior of the system is perhaps Chaotic. But if the individual nodes reach some stable state, even one in which disparate views of the "system" cause it to partition into two, the system's behavior is Complex:

A set of anticipated, linear, and deliberate behavior patterns at the microlevel are in fact combining under unanticipated conditions to produce unanticipated results at the macrolevel.

When a combination of actual events stimulates recovery paths in different nodes that then collide, the system's behavior is unanticipated, but emergent in the sense that it is an extension of the decision capabilities built into the application.

While some recovery paths are determined by the development team's code, others are not – not because the code flows are "wrong" or "defective" but because the development team did not *envision* and did not write code to *respond* to all eventualities. The software system is exhibiting its resilience in spite of its developers' lack of foresight. The software system's behavior is limited by the encoded recovery mechanisms, but remains free or unconstrained because the mechanisms do not address all possible eventualities.

To sum up, the Complexity of the software system ultimately derives from the projections and utterances of the development team as they are realized and combined in the execution event. The development team projects a trajectory for their code flows and for the product behavior; but they do so from a horizon circumscribed by their expectations and their desires and their experience with how products like this behave.[42] What they express can be emergent in its effect or meaning if they have not circumscribed its executed-meaning, and/or if they have imprecisely specified or understood its reception in or reaction to its execution environment. They might not ever notice this Complex behavior unless during execution the right sequence of events and responses occurs. But as I pointed out earlier, Complex systems are Complex regardless of whether we recognize them as such at any given moment. They are emergent in two senses: 1) our recognition of Complexity is revelatory, appearing suddenly in place of non-Complex behavior; and 2), Complex behavior is paradoxically both limited and enabled by the existing structural potentialities in the Complex entity.

COMPLEXITY: MISTAKE, LIMITATION, OR INVITATION?

I anticipate two objections at this point: a) this is not Complex behavior, it is bad design and/or bad coding, and b) this is not Complex behavior, it is lack of foresight. The first objection confuses Complexity with the value we place on a new state of a system.[43] The software system has adapted to unforeseen events and it continues to function. Neither users nor developers may want the system to behave this way. The new state the system has achieved may not meet our expectations and it may be of far less value to us than its normal state. But neither Complexity nor the "order" in Chaos was ever required by Design to serve utilitarian ends.

The second objection actually supports my claim that the system behavior is Complex by virtue of the phenomenological context in which it was created. Just as we recognize a "bad" play after the fact, we can certainly play Monday morning quarterback to the development team. But to do so ignores the role of the execution environment and the innate historicity of our understanding. Suggesting that the development team who wrote the Complex code was somehow inept is like taking Newton to task because he wasn't Einstein. To foresee Complex behavior, in the special case we are examining here, would be to eradicate it. The development team would have to have anticipated and projected their encoded utterances into a context comprising a myriad of 'initial conditions' and interdependencies:

- The range of values for any status, data field, or control field
- The range of possible protocol responses (valid or invalid)
- The possible interrelationships between all code paths, including decision points 'downstream' from any given decision point and their interdependencies
- The effect of elapsed time between operations, protocol exchanges, and recovery operations, and what effects resource wait states might have
- The system's sensitivity to execution time
- The system's sensitivity to combinations of paths executing in parallel
- The number of paths in the system and the complexity within those paths
- The kinds of transitional situations (failure modes) and how these can be stimulated
- The effect of other characteristics in the system's executing context (such as number of errors encountered, the depth of error handling, the ability to roll back from any point in any path independent of all other paths, the time of error detection and the decision ability of the detecting paths, etc.)

Is such understanding impossible? No, not in theory. Aren't there tools and techniques already available to the industry that help development teams see 'the big picture'? Certainly there are. I am no more trying to excuse miscreant development and defective product than I am trying to present

Complexity as the software industry's Pre-Determined Doom. The point here is that for any given set of encoded utterances that does not arise from a horizon in which all these factors (and many others not mentioned here) are 'visible', there is the potential for Complex behavior under the right conditions. In such cases there is a surplus of meaning (to appropriate Ricoeur's phrase): what the development team intended their code to mean/do/effect is less than what it can mean/do/effect under some circumstances.

Object oriented development methodologies, reuse, certain test methodologies, and development discipline such as performing inspections are often touted as effective approaches to software development. Based on what we know about Complexity, there's no obvious reason why these techniques can't give visibility to the unforeseen, why they can't help contain the possibilities of projected utterances. But we also are now better positioned to understand the challenge to software development and quality assurance described in chapter 2. The special case of Complexity we have examined here is accessible only through the experiential strategy; yet accuracy and precision, which are crucial to justifying one method rather than another, are (at least on the surface) coarse grained endeavors. Drawing conclusions about an approach's accuracy based solely on the incidence of defects ignores critical factors in the genesis of defects. And one cannot verify an approach's precision based solely on 'objective' data unless one can also show compatibility with the structures of experience.

The customers of the software industry would certainly like to see this special kind of Complexity eradicated, but doing so may not be feasible, either methodologically or financially. Time to market may prevent the decomposition of requirements into a formal metalanguage that limits the effect of individual developer's horizons. The characteristics of Chaos and the multiple nested systems that surround the developer's cubicle indicate that even transcription may not prevent errors in the code or in the executing system that result in Chaotic or Complex behavior. But in one crucial aspect, the nature of Complexity and Chaos in software may be an invitation rather than a limitation. Just as "things" are a cooperative effort between agents, structures, and events, and just as Complex systems are the interplay of perception, systems, and time, so perhaps can our recognition of Complexity in software help guide us to a better understanding of what role we can and cannot play in "the mystery" of Complex but utilitarian artifacts, especially software itself.

This last statement may seem circular, perhaps even starry-eyed. But in fact it is congruent with what we know of Complex systems. They evolve through the interaction of components, just as societies change through societally dispersed changes whose "causes", for convenience, we like to assign to one or more highly visible individuals. As Paul Cilliers has said, we must know the history of a Complex system in order to fully understand it, in part because the system's history helps determine the system's structure (Cilliers 1998, 107ff). If software is a projection of our human understanding, rather than a thing or an object, then what we learn "about" software teaches us something about ourselves, especially our limitations with respect to software production.

Looked at more pragmatically, perhaps it is time to stop assuming software can be useful in all circumstances. One need not take the view of Neville Holmes and others that software and computers have complicated rather than simplified our lives, exacerbating the very problems they were intended to solve (Holmes 1998). But it may be time to realize that "If you design it, they will come" is no longer credible from a professional or financial viewpoint, especially if the product being developed may be prone to Chaotic or Complex behavior. The industry must define its limitations with respect to Chaos or Complexity, drawing boundaries around itself within which it can pursue classically inclined engineering and beyond which it can make only limited claims about product quality.

Much work remains before the industry can draw such boundaries. I have identified only a few characteristics of software products that may produce Chaotic behavior. There are no doubt others, such as the relationship between processor speed or chip design and code structure. There are other arenas to explore – where, for example, is the upper limit for successful scalability, and what factors contribute to Chaotic behavior once that limit has been reached? Computational theorists may well identify deterministic structures in code syntax and semantics that will prohibit certain kinds of potentially Complex algorithms or promote tried-and-true patterns to the status of formally justified engineering principles. In any case, the new science has sounded the challenge to the software industry, and if my analyses here are even remotely correct, the industry must confront a kind of Chaos and Complexity that promises to answer that challenge with a few of its own.

EPILOGUE:
THE SCIENCE OF COMPLEXITY, THE COMPLEXITY OF UNDERSTANDING

> Things and beings in the Time order – even when to all appearance complete, as a body is when fit to harbour a soul - are still bound to sequence; they are deficient to the extent of that thing, Time, which they need: let them have it, present to them and running side by side with them, and they are by that very fact incomplete; completeness is attributed to them only by an accident of language.
>
> Plotinus, *The Enneads*, III, 7, 6

Time. There's always too little of it when we are enjoying ourselves and far too much of it when we are bored. We linger in it in reverie, struggle against it in traffic. It imbues everything we are, from the circadian rhythms of our physical body to the concepts and memories that comprise our sense of self.

Whether time is real or a construct of our minds, whether it is reversible or not, even whether time itself had a beginning, are important questions for philosophers, theologians, and scientists. And time, in the form of emergence, plays a crucial role in Complex systems. An emerging system may well look, at any particular moment, as if it were "complete" – its future evolution may lie masked within its present laws and patterns of behavior. By rights, we should have a complete understanding of Time before we could completely understand Complexity. We may never have that, and we must be content to run side by side with Time, knowing that what we divine about Complex systems today may not be adequate tomorrow.

That is less a limitation than a feature, as we like to say in software, of emergent entities. We come to know them incrementally; they exist as emergent systems only in time, and our desire to fathom them is fulfilled only in time. Radically revising our views of the physical and spiritual worlds in which we live, emergent Complex systems have taught us that how we understand ourselves and the world around us depends more on time than on our perspicacity. The recent history of science, especially physics and cosmology, suggests that this constraint also applies to some of our most dearly prized 'objective' knowledge, particularly our knowledge of the universe and our place in it (Rees 1997, Penrose 1996, Prigogine 1996). If the historians of consciousness like Allan Combs are to be believed, even our awareness of ourselves and of the world has evolved in ways we understand only with hindsight (Combs 1996). It is as if it were humankind's fate to always be at the periphery of knowledge, getting merely a glimpse of the full potential of what we comprehend as it streams away into a future we cannot penetrate.

It is not surprising that we find ourselves at the mercy of time. Time and the space in which it reveals itself are the fundamental constituents of the horizon described in chapter 6. The horizon is always more than the particular object revealed within it, providing an excess of significance to the thoughts that our understanding situates within it. Unseen as such, time and space are, like the horizon they inform, the precondition, and promise, of thought. When we encounter a Complex system, we situate it, snapshot it at a stage in its evolution as well as ours. This historicity, this being bound to sequence as Plotinus calls it, is for many Complex, emergent systems their most problematic characteristic.

Of course, because time is one of the primary colors that contribute to our palette of observations, thoughts, and knowledge, this historicity isn't always apparent. After the preceding discussions of strategies one would expect that this historicity might not be as evident when using the coarse grained strategy as it might be when using the experiential strategy. Given what we've learned about the interplay between observer and observed entity, it should come as no surprise that the "object" of study exerts as much of an influence as the strategy used to study it. The degree to which our understanding of a system is influenced by our historical and horizonal "situations" may differ depending upon whether we are examining rules or events, seemingly inanimate entities or human artifacts. Two examples will help illustrate these points.

In cellular automata, an evolutionary sequence of state changes is created for a bounded system. The state space is divided into cells, each with a set of possible states (e.g., "on" or "off", "populated" or "empty", shows a number from 1 to 6, etc.). The collection of cells undergoes evolution over specific time slices in accordance with some generative rule, such as "With each generation, the fourth cell from the cell most recently turned 'on' also changes to the 'on' state." Even with such simple systems and generative rules, complete state transitions can occur – such as a system starting with some cells 'on' ends up with no cells 'on'. With the generative rule in hand we can manipulate the system however we please, and while any given set of generated system states may not exhaust all possible states, the states we generated are intelligible because we know the generative rule. In other words, the rule can be used to explain the behavior of the system (Peak 1994, 302ff).

Of course, in this case the deck is stacked decidedly in our favor. We chose the rule. We even chose the objects and the characteristics to which the rule applies. But with a little more sophistication in the definitions of cells, their states, and the rules they follow, one might model population migrations or even primordial gene development (Tsonis 1997). A population model and its rules might be verifiable through experimentation. On the other hand, an evolutionary model for a system that no longer exists could not be verified; it could only be found coherent with additional data about the system. But in either case, the "ordering principles" of the system as analyzed are mathematical and logical abstractions, and the strategy employed is obviously coarse grained. In these cases, our understanding of the system is not time bound (with some caveats to be listed later). Given the constraints built into the system by the strategy and the definitions of cells, states, and rules, it doesn't matter *when* one examines the system. As is the case with primordial genetic evolution, it may not even matter *if* one can, literally, visually, examine it at all. Time and the algorithms of the system's evolution march forward with or without us to cheer them along.

But that's not always the case. Our second example comes from the same discipline, molecular biology, but illustrates a very different view of Complexity and highlights two issues germane to examining events and artifacts: order and value. Combs and Goerner cite cellular evolution as an example of a Complex system. On the account they cite, a more complex organism was created when the parasitic precursor to modern mitochondria infested the ancestors of its modern day eucaryotic host (Goerner 1995, 28; Combs 1996, 37). This new cell, the likes of which you studied in high school biology, is considered to be a "higher order" entity than its constituent parts.

And not just in a taxonomical sense: the new cell was more efficient and better adapted to development and to differentiation to meet the demands of its environment. On the face of it, we seem to have another 'objective' instance of Complexity – two entities merge and a third distinct entity emerges. But there is a crucial difference between these two examples.

First, the emergence of this Complex entity is not a deterministic event at least in any significant sense: one would have to stretch credibility to say we could have predicted the original infection. Second, the evolution from the non-Complex states of the constituent parts to the Complex state of the eucaryotic cell is not described by a rule *per se*. It is described as an event that creates a new structural and functional relationship. The eucaryotic cell is physically different from its constituent parts, and the biochemical relationships between those parts and each other (or between the parts and the extra-cellular world) are different. But note that there may be interpretive prejudices lurking in what appear to be raw descriptions: "higher", "order", and "adapted" may be honorific titles or they may be a value-free observational lingo. In either case, it is clear that Complexity can be recognized by rules or by structural or functional novelty. The rule allows us to project future behavior; structural or functional evolution is recognized only *post facto*.

"Novelty" brings us right back to the historicity of emergence. For systems whose rules are not (or cannot be) known, Complexity is comparative, it is something new compared to what preceded it. But novelty alone is insufficient grounds for calling something Complex. If we cannot understand structural relationships or functional relationships between the precursors and the ensuing system, we cannot legitimately call the ensuing system Complex. We encountered a similar problem in the discussion of Chaos. Simply supplying random inputs to a software program was not an exercise in Chaos unless and until we could specify how those inputs functioned as initial conditions for the Chaotic system. "Garbage in, garbage out" is not one of the laws of Chaos. And as we saw in the case of the termination of the spreadsheet program, unanticipated behavior isn't necessarily Chaotic or Complex.

So when we discover a new state, a higher order, in a Complex evolving system, we must be able to discern how the new structure was "latent" in the old one. I am not suggesting that all structural configurations of a system are in some metaphysical sense omnipresent even though only one is available to our gaze at a given time. Complexity isn't some teleology verifiable only in time and through time simply because we view it 'through a glass, darkly'. Like the archeological history of Stonehenge, we must understand how what came before affected what ensued. In the case of rule-based systems, we face only constraints on our knowledge:

- We may not know all the relevant relationships in or characteristics of a system, and with the limited knowledge at our disposal we misproject the system's behavior.
- If the system's evolution does in fact follow some generative rule or some structural pattern, neither may be accessible to us at the moment we view the system. As a result the system's behavior may appear to be random or indeterminate in its broadest sense.
- If the system's evolution does in fact follow some generative rule or some structural pattern, neither may be accessible to us *no matter when* we view the system. As a result the system's behavior may appear to be random and inexplicable.

But for artifacts, especially linguistic artifacts like software, we must appreciate the role our own horizons play in constituting *both the entity prior to emergence and the emerging entity*. Artifacts aren't like gravity: with artifacts you can take them or leave them, but most of us will have to make our peace with flights of stairs and bathroom scales whether we like gravity or not. When the system being studied is an artifact, it seldom escapes its purposive past or its purposive present. It's good for something, it meets expectations; the software does its job, the reorganization is good for the business. Artifactual systems appear in the world as intentional creations that, to paraphrase Wordsworth's *Tintern Abbey*, come trailing clouds of beneficence behind them. "Complexity" in such cases is likely to be a "cooperative effort" between things, observers, and structures. As such, our understanding of the system may well evolve as much as the system itself does (although not necessarily at the exact same time).

A "science of Complexity" for artifacts will need to be significantly different from a science of atomic particles. Even if one takes the latter as theoretical constructs with no real-world referents, the only intentional, symbolic, and semiotic factors in quantum theory are in the theory itself. With artifacts, the "object" of study is intentional, symbolic, and semiotic; as a result, any science of Complex inanimate objects must give way to a science that accommodates the "Complexity of understanding". Time in the case of artifacts is different: it is interpersonal, historical, and experiential. For example, if we are to understand an historical artifact, *as* an historical artifact and not just some quaint dumb nick knack, something has to bind our horizon to that of the artifact's creator. We must see in it the traces of its purpose, of the meaning it holds in proxy for the craftsperson who cannot speak directly to us. The artifact is present to us, its meaning seems to bridge the past and the present, and that past is someone else's, not ours. The problem of historical understanding is obviously tied closely to the problems of the nature of time and space and of the nature of consciousness. Perhaps one day we will understand how in conscious thought we can overcome limitations that have doomed time machines to science fiction film archives.

In the meantime, even the current "science of Complexity" seems to be diverging from the old subject-object paradigm. There's much debate about particulars, but there does seem to be a growing consensus about at least the cooperative relationship between mind and world, present and past, based on the innate, formative (and *in*formative) structures which indwell both.

Vandervert suggests mathematics is one such structuring principle that originates in neural patterns. The success mathematics has had in describing "reality" is due to the mutual and cooperative emergence of things and mind (Vandervert 1993 & 1994). At the end of *Shadows of the Mind*, Penrose explores the nexus of the physical, mental, and mathematical worlds. He does not embrace anything as explicit as Vendervert's position, but he approaches something like it when he denies that mathematics occupies an ideal Platonic realm and he denies it is a mode of thought only (Penrose 1996).

Allan Combs, while he agrees with Vandervert that mind (consciousness) and reality are coeffecting, does not share Vandervert's view that mathematics is the only language in which the fundamental principles of reality can manifest themselves to thought (Combs 1996). And Paul Cilliers presents us with a world in which scientific and non-scientific knowledge are simply different narrative forms for telling the story of the flow of experienced time, where any "individual" is simply a linguistic tag for the nexus of the interrelationships that are, for example, "biochemist", "mother", and "women's basketball coach" (Cillier 1998).

Scientific language – be it theoretical primitives or observational language – is not immune from social and historical change any more than it is unaffected by the structures of experience operating in those who use the scientific language. As the linguistic system changes, as the social system evolves, as new vistas on "reality" are opened up, science will also change. Complexity is not the end of science, it is not a limit to knowledge; on the contrary, it admonishes us that "end" and "limit" may be a now-outmoded vocabulary. Like other theoretical jargon, it will need to be refined and assimilated into the new paradigm (Durlauf 1997).

WE HAVE MET COMPLEXITY, AND IT IS US

In previous chapters we have explored some of the structuring principles of experience and their interplay in horizons and understanding. Such do not provide Truth; they provide coherence. Ideas and perceptions fit together in our experience, they appear to us whole and unmediated precisely because they appear to us within structures of thought and perception. They are constituted as ideas and perceptions by those structures or principles. Whether the principle is a geometric axiom or our incarnate understanding of the unseen sides of a dwelling, we are co-architects of the world around us.

So when we encounter order, what is it? One might assume it is a correspondence or correlation between patterns of thought and perception and patterns in what is perceived and understood. While we might all have our "private" windows on Chaotic and Complex systems, in order to share our discovery, our understanding of the system must be made symbolic, linguistic, shareable. Whether that language is mathematics or geometry or metaphor or propositions, it is the linguistic expression that gives our encounter with the system durability and exposes it to reflection and analysis.

And once that analysis has begun, the system changes for us. Our concept of Chaos has radically changed what "unpredictable weather" means to us - it is just too complex to model and control, it's not that it is merely capricious. Our understanding of emergent systems has helped us appreciate the reasons business organizations seem unruly, even if we don't yet know by what rules they can be controlled. Once a Complex system is recognized as Complex, our perceptions of the system are forever changed. Our horizon expands to include it; the context in which we understand that system, and perhaps others, remains forever perturbed by the dynamics we discovered when we went hunting for linearity.

Prigogine speaks of a "buildup" of correlations of particles that produce memory effects that in turn produce time irreversibility at the macrolevel. Individual particle trajectories combine and the new collective trajectory represents the realization of one possible trajectory for each particle in the collective. Correlation collapses possibility into actuality at the level of collections of particles just as (on some interpretations) measuring a particle realizes it in some space and time. But once committed, there is no turning back. Like a gene, each particle's current trajectory is a kind of information that will propagate into the future: each particle's trajectory influences and is influenced by the collective and affects any other particle that it encounters.

We might metaphorically extend that notion to say that our encounter with the "order" in Chaotic systems or the "generative principles" of Complex systems is a correlation of the system with the observer. When we encounter Chaotic and Complex systems we actually encounter a particular realization, an instantiation, of one possible outcome for the system. We are witnesses to a moment in the system's "history", yet we also participate in that historical event by virtue of our presence. "Chaos" and "Complexity" are essentially the effects of a situated, historical, or "localized" understanding. Systems don't suddenly become Chaotic or Complex, any more than raging rivers and weather patterns behave differently today than they did a million years ago. But today they have been encountered and understood as Chaotic or Complex. When systems are promoted from chaotic or complicated to Chaotic or Complex, they leave a legacy. We cannot forget the discovery of Complexity in a system any more than we can reverse our internal, experiential sense of the flow of time. The concept of the horizon of perception and understanding shows that in some ways "we" are what we have learned. As Sally Goerner puts it, "the mystery is in us, of us, and more than us" (Goerner 1995, 36).

ENDNOTES

[1] Newton can't take all the credit for the paradigm shift we now think of as 'classical' science. Margaret Osler provides an excellent overview of the philosophy of science up to and beyond Newton in Osler 1994.

[2] Outside of the physical sciences, these forays into Chaos have been well received but not necessarily cogent or complete. Wheatley 1992 assumes her rhetorical leap from quantum mechanics and complexity to business management needs no formal or empirical substantiation. Freedman 1992 is more cautious about the intricacies of Chaos but he strains credibility when he arbitrarily links Peter Senge's concepts to those of Chaos and Complexity. Olson 1993 conflates feedback with Chaotic sensitive dependence on initial conditions and nonlinearity in his discussion of Chaos and software development. And even in aesthetics, where one might expect more thorough analysis, Arnheim 1996 appears to reduce Chaos in representative art to conflicting impressions of the work, a credible position but one that requires more defense than Arnhiem's assertion that the "unity" of Chaotic art is derived from the gestalt of perception. We will examine such 'unifying principles' as they relate to software development in some detail in chapter 6.

[3] In fact, Chaos and Complexity are often cited in conjunction with a move *away* from the industry's current obsession with process control and *towards* people management - see Bollinger 1997 and Bach 1995.

[4] Even when comparative analyses are made, they often are little more than matching concepts from the compared approaches (e.g., Bøttcher 1997) or choose a few characteristics for comparison.

[5] For example, Heady and Smith in their 1995 "empirical analysis" tried to elucidate the difference between Total Quality Management and Quality Management by examining their incidence in articles. Their statistically justified margin of error does not account for the fact that some of their data was obtained

through semiotic analysis, which is a hermeneutic, not an empirical, technique of analysis. More recently, Seaman and Basili (1998) claimed data obtained through participative observation and structured interviews was "empirical", even though the hermeneutic effort in this case is even more significant than in the previous example. One is left wondering if in some circles quantifiable and empirical aren't considered synonymous.

[6] I do not want to get entangled in debates about specific approaches. I'm looking for a methodological justification for choosing one approach rather than another. The relative merits of ISO versus CMM, or of using function points rather than lines of code, etc. are interesting topics in their own right but such discussions are legitimate only after the methodology is in place.

[7] There are exceptions, but on the whole the industry seldom approaches the rigor of, for example, Beizer's analysis of Cleanroom (Beizer 1995b).

[8] Downs and Garrone claim 100 reliability models were proposed in a 15 year period, none of which survived either use or scrutiny (Downs 1991). Many writers have pointed out that reliability models need more than mathematical correctness (Brocklehurst 1990, Halverson 1992, Karama 1991, Lanubile 1996, Yang 1995). In the parlance used here, such models also need to identify the environments in which they are correct and the initial conditions that they treat as significant.

[9] If that seems like common sense to many of you, I should point out that the journals and conference circuits in software development and software quality are full of single instance "case studies" that are passed off as, or interpreted by the audience to be, 'methodologies'. And despite the fact that different kinds of software development organizations face different challenges and risks (see, e.g., Jones 1994), one still finds professional papers, tutorials, and seminars that present "one size fits all" approaches.

[10] Some would have us believe that messy collections are indeed the necessary and sufficient conditions for calling something Chaotic. For example, see Raccoon 1995a & 1995b.

[11] Whether there is 'downward causation' in hierarchies like this is a contentious topic we examine in a later chapter.

[12] In an irony best appreciated by those familiar with phenomenology's struggle to make a place for itself next to modern science, Chaos Theory has forced at least sociology to adopt a pluralism that exemplifies what phenomenology calls the "thematic" nature of understanding. We'll see this pluralism at work in this chapter, and explore the phenomenological view of understanding in chapter 6.

[13] This is, of course, the ideal situation. In practice, many theories lack precision for all instances of the events they intend to explain. But even in the class of "approximately true" theories, Chaos models remain especially problematic.

After discussing the various meanings of "approximately true" as applied to Chaos models or theories, Peter Smith is forced to conclude that such theories are approximately true if their geometric structure corresponds to aspects of the actual system "in respects the theory cares about" (Smith 1998, 269). The truth of a Chaos model or theory is "interest-relative" (271), a position Smith does not explore in detail but we will in this and later chapters.

[14] See, for example, Abraham 1995 vs. Stapp 1993.

[15] Strictly speaking accuracy and precision have in the philosophy of science different meanings than are used here, but I think the summation captures the spirit if not the letter.

[16] Sivak may believe Olson already proved that in his 1993 book *Exploiting Chaos*, which he cites.

[17] I am calling this type of justification 'formal' to avoid making it appear to be just another reduction of one discipline to another. While such justifications in software will need to draw from the disciplines of psychology or sociology or even physics or philosophy, the 'software specific event' is not to be 'explained' by reference to the appropriate statements in those other disciplines. I am suggesting that formal relationships between disciplines reflect the "extrinsic explanatory power, or relatedness" Coles speaks of with respect to cosmological theories (Coles 1997, 11) in physics. Software professionals create products, they are not social scientists or philosophers or scientists. But they are consumers of philosophical speculations and of research in the physical and social sciences. They assume that these disciplines have in fact identified 'types' or 'classes' or 'forms' of behavior that software professionals can use in analyzing their 'objects of study'. A formal justification in software thus appropriates some finding in some other discipline and hopefully shows that said finding is complete, consistent, and coherent when applied outside its original theoretical realm. On good days, it is a symbiotic relationship, where each participant benefits.

[18] Even the broad perspective I am assuming here might be limiting: it doesn't consider the physical environment in which decisions are made. James Bach tells of a case in which 200 defects was considered an excessive number when a marker board was used for tracking, but after an automated tracking system was installed 2000 defects caused no concern (Bach 1998).

[19] For some readers this position may sound too relativistic, but it *is* being recognized as a legitimate response to the dilemma faced by the soft sciences when they compare their methods and objects of study with those of physics. Mariam Thalos describes the distinction between the social and physical sciences as ultimately a metaphysical distinction: the social sciences often posit layers or strata of reality (what we have been referring to here as nested systems). This

position becomes problematic if the same notion of causation is applied to all levels or strata, because whatever affect higher strata may have on lower ones is a different kind of causation than operates in the physical realm. Structural explanation, whereby events in one system are understood by recourse to another more encompassing system, indicates that current notions of causation are inadequate (Thalos 1998, 291). Eventually, the 'stratified' view of human behavior and artifacts will need to define causation within its domain. For our purposes here, we'll take the common-sense view that nested systems do influence each other although we don't know the causal mechanisms with precision.

[20] I'm assuming math is a language albeit more consistent across cultures than the language codified in each culture's dictionaries.

[21] Both the coarse grained and the predictive or statistical strategies look at their objects of study the way test engineers often view software – as a black box. And both can be applied in the hope of making predictions about the future state of a system. I have chosen not to use "predictive" because in a significant sense, all efforts to understand systems (software or otherwise) have a predictive intent. Harvey and Reed are certainly aware of this, but I believe they can maintain their distinction between the predictive strategy and other strategies because of the logical requirements for scientific explanation and theorizing. That regimen is lacking in the software industry, so calling this strategy "predictive" would be potentially misleading.

[22] Such defect typologies are rare. The industry seems to be content with High/Medium/Low priorities and some variation on the Critical/Severe/Important effect scheme. However, see Bridge 1997 and Kelsey 1995b.

[23] I say "traditional" because in the next chapter we will in fact call upon phenomenology, a discipline some might consider to be a form of psychology, to help us describe some of the limitations of coding and the impact of those limits on the program's execution.

[24] In his review of Price's *Time's Arrow and Archimedes' Point*, Callender points out that physics (Chaotic or otherwise) is time biased in that it typically describes lawful dynamics based on initial conditions, neglecting lawful behaviors based on final conditions or a mixture of initial and final conditions (Callender 1998, 151). Although he is concerned with more general issues in time reversibility, his point is well taken here in that in software Chaotic systems we probably have to consider both initial and final conditions as part of the parameters.

[25] There are of course models for parts of the execution (e.g., routing and network performance models), but to my knowledge no one has modeled the complete system of execution in the environment we will be exploring. For the purposes of this discussion, it is simply inconvenient that we have no models today to work

[26] with; the structural approach will still allow us to make some progress in defining Chaotic execution.

[26] One could object that the data exchange programs on desktops A and B and the protocols used between them should have been made more robust. Any software engineer with a moderate amount of paranoia would have designed the programs and protocols to handle such cases. In other words, the developer should have anticipated this failure scenario and designed the code to handle it. I have never claimed that Chaotic behavior was unavoidable, but we do have to understand what Chaotic behavior is before we can try to avoid it. "Good programming practice" or not, the example serves to illustrate how Chaotic behavior can occur in software execution.

[27] The developer's intentional creation of coded statements and directives is one reason why software and physics part ways forever when it comes to Chaos. In physics, "Things Happen" without intentions or expectations getting in the way. But software development intends to create what will happen; it constrains future possibilities by creating the code that will perform specific functions. Of course, the execution environment adds additional complexity, and software execution is where whatever *can* really happen, really does happen. Deficiencies in the process of creation may be reflected in Chaotic behavior 'downstream', or behavior downstream can reflect Chaotic possibilities no one ever imagined during development. Either way, there is Chaotic behavior that needs analysis.

[28] I am aware I have opened a philosophical can of worms here. When I introduced purposiveness or anticipated outcomes, I introduced intentionality into the definition of Chaos and ultimately into the definition of Complexity in software. That leaves my arguments here (and those in chapter 6) open to a fairly standard objection from the proponents of computational theory of mind (or from those pursuing mathematical modeling for software engineering). The current research in artificial intelligence and computational theory of mind, so the objection goes, has been successful enough even on small scales to indicate that there's no need to introduce psychic events into models of intelligence, language, or, here, software execution. Even if Chaotic, such systems are adequately and completely described by mathematical means. In theory, we should be able to create a computational model for software execution, especially since the model and the entity being modeled are cut from the same cloth.
In responding to this objection, it is not enough to point out that we don't currently have such a model, and so the objection is based on faith not fact. The objection could still be true regardless of the state of our current modeling efforts. John Searle's famous Chinese Room argument won't help us here, either. Although Searle is sometimes interpreted as having freed intentional thought

from being reduced to computational action (e.g. Combs 1996, 20), others have offered strong arguments against him (see Cilliers 1998, chapter 4, for an overview of these arguments). Roger Penrose suggests that at best Searle merely distinguished actual events from their simulation (Penrose 1996, 40-41). Nothing says the simulation can't be accurate and explanatory. If we could model Chaotic software execution without recourse to expectations or intentions, the model would *function* as if true regardless of the presence or absence of intentions. And we know that, using a coarse grained strategy, we could easily ignore affective components of software development and execution and probably someday model Chaos in the software-as-executing-in-some-environment.

Penrose has addressed this kind of objection specifically in the context of theory of mind. He suggests that (physical) Chaotic models are computational and they are correct in that if they cannot predict the actual outcome of the system, they can still predict a typical outcome (23). But he also musters a detailed argument against reducing consciousness to computation. While the relationships between mind and its simulation, and executing software and its simulation, are certainly not equivalent, Penrose's argument suggests that for some entities, intentionality (he calls it "awareness") cannot be ignored.

So it is not a question of "making room" for intentionality in software Chaos or Complexity. Rather, the issue is what we must ignore if we do not allow explorations of intentionality as part of the tool kit. We can accept any success the coarse grained strategy may have, but we must also recognize that such success does not exclude other models or explanations, it does not exclude other factors and other systemic relationships.

Still, we have not yet established how intentions influence the software that is executing. After all, if there is no 'sphere of influence' that unites developer intentions with the code executing in real time, then there's no need to worry about reduction. But unlike physical systems, software is an artifact of human effort, and like the written text it shows the legacy of that cognitive and experiential context in which, and through which, it was created. The next two chapters will explore the relationship between intentions, implementations, and execution.

[29] It is worth pointing out that the arguments against reductionist approaches don't come just from the phenomenological camp. See Penrose 1996, Horst 1996.

[30] I am here adapting Gadamer's concept of play. For his own treatment, see Gadamer 1994, Part 1, section II.

[31] Penrose's goal is to find a biophysical basis for consciousness that exceeds analysis by computational theory. He proposes that a new physics is required to explain the origin of consciousness, a physics that understands the 'borderline' between classical and quantum formulations of natural objects. He specifically

suggests that the interplay of deterministic quantum events in neural microtubules, when sufficiently entangled with the local environment, will produce the variability required for the non-computationally-deterministic event of 'free' conscious thought.

This is not a description of the process of thinking, of course. It merely challenges the classical mechanistic view of mind on its own ground as it were. In what follows, I will be using phenomenology to describe some structuring principles of thought. While this differs significantly from Penrose's effort, I do believe that the phenomenological analysis presented here supports his goal of freeing thought from the constraining matrices of computer automata models.

Penrose himself might object to the discipline I am using 'in his defense' – historically there has been some tension between the hard sciences and phenomenology. But I think there are enough intriguing similarities between "the event of utterance" and the various notions of "the quantum event" to perhaps excuse my crossing the lines of battle. At the very least, both are actualizations of one of many possibilities and for both the mechanisms of actualization and the predictability of any given actualization are hotly contended.

[32] At the risk of irritating proponents of other schools of thought about language, I don't plan to defend the phenomenological view taken here. For the event of software coding, my assertion that language structures experience and understanding is reasonable regardless of whether one believes language reveals Being or one believes it is merely a game of conventions. By the time we examine Complexity in software in chapter 7, however, differences between the various schools will become more significant. For a different treatment of language, Complexity, and science, which might also be adapted to the analysis of Complexity in software, see *Complexity and Postmodernism* (Cilliers 1998).

[33] The example of the house receives more detailed treatment in Kockelmans 1966, 44ff, and Merleau-Ponty 1962, 67ff.

[34] Joseph Kockelmans provides a very accessible summary of the hermeneutical critiques of natural science in "On the Hermeneutical Dimensions of the Natural Sciences" in Kockelmans 1993, 100-126. See as well Heelan 1991.

[35] If you would like to experience these preconceptions a little 'closer to home', grab your company's quality manual, take the stairs to the roof, and drop the manual over the edge of the building. Making adjustments for its lack of aerodynamics, it will fall 'straight' down to the ground. You've just 'proved' that the Earth is stationary. If the Earth were rotating, the Earth-bound building would have been rotated away from the free falling manual. You 'know', of course, that the Earth rotates, and the 'visual' evidence doesn't dissuade you from that

knowledge. Feyerabend 1986, chapters 6 through 9, explores the relationship between naïve realism and observational language in detail.

[36] The notion of historicity, under different terms, appears even in discussions of predictability within physical and Complex systems. See Gell-Mann 1997 for a discussion of probability and indeterminism at the intersection of cosmology, quantum physics, and Chaos.

[37] In his analysis of Copenhagen and Bohmian quantum theory, Barry Loewer points out a number of the difficulties in these two interpretations of quantum events. The indeterminacy of the Copenhagen school is replaced by "irremediable" ignorance on the Bohmian interpretation. If the Copenhagen view makes events mind-dependent (measurement dependent), the Bohmian view leaves us uncertain whether to believe in its 'reality'. Resolving the issue of quantum indeterminism and determinism promises to only be the tip of the iceberg if a quantum approach to mind is to be successful.

[38] Perhaps it is time to apply a Complexity-oriented analysis of some features of Gadamer's problematic concept of *wirkungsgeschichtliches Bewusstsein*, the "effective historical consciousness" that allows us to understand historical utterances as both historically mediated and expressively immediate.

[39] I think a strong case can be made that this form of Complexity is the genus, and the Complexity of artificial intelligence software is a special case, not vice versa. But since AI applications are less controversially Complex, I am content to present them as the simple exemplar of Complex software behavior.

[40] Some might find this association of the developer's intentions and the executed code to be too simplistic in that it ignores a number of material and semiotic issues. I suggest that such issues are irrelevant from the phenomenological, experiential perspective we are taking here. I may no more care about (I may be no more conscious of) the mechanisms of computer processors when I write software than I care about (or am aware of) the physics of sound and the biological structures I use when I speak. Steven Horst addresses some of these 'intermediary' issues in Horst 1996. Particularly poignant is his analysis of the functional and semiotic aspects of executed code: while it is obvious that the electrical events in the processor chip are physically independent of the developer's mental state, computer programming is successful to the degree that the physical computational event expresses (performs, instantiates) the developer's intention (Horst 1996, 129ff).

[41] There's nothing mystical about this identification. Pick up any novel that employs an "omniscient narrator", and you will personally experience this same sort of identification and participation. As your eyes scan the ink on the page, a voice emerges in your understanding, an entity or personality who is not you yet is of

you, an entity or personality that is separate from, yet is able to merge seamlessly with, the agents in the plot.

[42] One might imply from this that maintenance engineers should never be organizationally separated from development engineers. Making developers responsible for maintenance completes the learning curve with respect to the software they produced. As the product reveals its "features" over its life span, the information about actual behavior should be cycled back into the development team so that it can become part of their context when they create other products.

[43] The concepts of teleology and evolution are sometimes difficult to separate, and this seems especially true for studies of Complexity in the social sciences. The risk is greater when discussing software, which is intrinsically seen from a utilitarian, value-laden perspective. Like strands of DNA, the intentions and expectations inherent in the code distributed across the nodes in our example can, in time and under certain circumstances, combine 'as usual' or recombine to produce new behavior. To echo a point made in chapter 3, "things happen", whether we appreciate the event or not.

BIBLIOGRAPHY

Abraham, F. D. 1995. "Dynamics, Bifurcation, Self-Organization, Chaos, Mind, Conflict, Insensitivity to Initial Conditions, Time, Unification, Diversity, Free Will, and Social Responsibility." In *Chaos Theory in Psychology and the Life Sciences*. Edited by R. Robertson and A. Combs. Mahwah, NJ: Lawrence Erlbaum Publishers. 155-73.

Achinstein, P. 1993. "The Pragmatic Character of Explanation." In *Explanation*. Edited by D-H. Ruben. New York: Oxford Univ. Press. 326-344.

Arnheim, R. 1996. "From Chaos to Wholeness." *Journal of Aesthetics and Art Criticism*, 54 (2): 117-120.

Aveni, A. 1997. *Stairways to the Stars*. New York: John Wiley & Sons.

Bach, J. 1995. "Enough About Process: What We Need Are Heroes." *IEEE Software*, 12 (2): 96-98.

Bach, J. 1998. "The Highs and Lows of Change Control." *IEEE Computer*, 31 (8): 113-15.

Bardyn, J., & D. Fitzgerald. 1997. "Chaos Theory and Project Management." *Proceedings of ProjectWorld 97*. Vol 1. Wash. DC: ProjectWorld Inc. E4/1-12.

Beizer, B. 1995a. *Black-Box Testing*. New York: John Wiley & Sons.

Beizer, B. 1995b. "The Cleanroom Process Model: A Critical Examination." *1995 Proceedings*. Pacific Northwest Software Quality Conference, Portland OR, 27-28 September 1995. 149-173.

Bollinger, T., & C. McGowan. 1991. "A Critical Look at Software Capability Evaluations." *IEEE Software*, 8 (4): 25-41.

Bollinger, T. 1997. "The Interplay of Art and Science in Software." *Computer*, 30 (10): 128, 125-27.

Booch, G. 1991. *Object Oriented Design*. Redwood City, CA: Benjamin/Cummings Publishing Co.

Bøttcher, P. S. 1997. "Comparing Total Quality Management and the Capability Maturity Model (CMM) in a Organizational Change Perspective." *Proceedings of the Seventh International Conference on Software Quality*, Montgomery AL, 6-7 October 1997. 67-86.

Bridge, N., & C. Miller. 1997. "Orthogonal Defect Classification using Defect Data to Improve Software Development." *Proceedings of the Seventh International Conference on Software Quality*, Montgomery AL, 6-7 October 1997. 197-213.

Brocklehurst, S., et. al. 1990. "Recalibrating Software Reliability Models." *IEEE Transactions on Software Engineering*, 16: 458-470.

Callender, C. 1998. "The View from No-when." *British Journal for the Philosophy of Science*, 49: 135-159.

Casey, E. S. 1976. *Imagining: A Phenomenological Study*. Bloomington: Indiana Univ. Press.

Chamberlain, L. 1995. "Strange Attractors in Patterns of Family Interaction." In *Chaos Theory in Psychology and the Life Sciences*. Edited by R. Robertson and A. Combs. Mahwah, NJ: Lawrence Erlbaum Publishers. 267-73.

Cilliers, P. 1998. *Complexity and Postmodernism*. New York: Routledge.

Coad, P., & E. Yourdon. 1991. *Object-Oriented Design*. Englewood Cliffs, NJ: Yourdon Press.

Cohen, L. 1991. *Inspection Moderator's Handbook*. Maynard, MA: Digital Equipment Corporation.

Coles, P., & G. F. R. Ellis. 1997. *Is the Universe Open or Closed?* Cambridge: Cambridge Univ. Press.

Combs, A. 1996. *The Radiance of Being*. St. Paul, MN: Paragon House.

DeGreene, K. B. 1997. "Field-Theoretic Framework for the Interpretation of the Evolution, Instability, Structural Change, and Management of Complex Systems." In *Chaos Theory in the Social Sciences*. Edited by L. D. Kiel and E. Elliot. Ann Arbor: Univ. of Michigan Press. 273-294.

Dobbins, J. H. 1987. "Inspections as an Up-Front Quality Technique." In *Handbook of Software Quality Assurance*. Edited by G. G. Schulmeyer & J. I. McManus. New York: Van Nostrand Reinhold. 137-147.

Downs, T., & P. Garrone. 1991. "Some New Models of Software Testing with Performance Comparisons." *IEEE Transactions on Reliability*, 40: 322-328.

Drabick, R. 1997a. "Modeling the Formal Testing Process: Conclusion." *Software QA*, 4 (5): 37-39. The series appears in 2(2), 3(1), 3(4), 3(5), 4(3), 4(4).

Drabick, R. 1997b. "Growth of Maturity in the Testing Process." *Software QA*, 4 (6): 46-48.

Durlauf, S. N. 1997. "Limits to Science or Limits to Epistemology?" *Complexity*, 2 (3): 31-37.

Ebert, C. 1997. "The Road to Maturity: Navigating Between Craft and Science." *IEEE Software*, 14 (6): 77-82.

Elliott, E., and L. D. Kiel. 1997. Introduction to *Chaos Theory in the Social Sciences*. Edited by L. D. Kiel and E. Elliot. Ann Arbor: Univ. of Michigan Press. 1-15.

Evans, J. R. 1996. "What Should Higher Education Be Teaching About Quality?" *Quality Progress*, 29 (8): 83-88.

Evans, M. W., & J. J. Marciniak. 1987. *Software Quality Assurance and Management*. New York: John Wiley & Sons.

Fenton, N. E. 1991. *Software Metrics: A Rigorous Approach*. New York: Chapman & Hall.

Feyerabend, P. 1986. *Against Method*. Thetford, England: Thetford Press.

Freedman, D. H. 1992. "Is Management Still a Science?" *Harvard Business Review*, Nov.-Dec.: 26-38.

Gadamer, H-G. 1977a. "Aesthetics and Hermeneutics." Translated by D. E. Linge. In *Philosophical Hermeneutics*. Edited by D. E. Linge. Berkeley: Univ. of California Press. 95-104.

Gadamer, H-G. 1977b. "Semantics and Hermeneutics." Translated by P. Christopher Smith. In *Philosophical Hermeneutics*. Edited by D. E. Linge. Berkeley: Univ. of California Press. 82-94.

Gadamer, H-G. 1977c. "The Universality of the Hermeneutical Problem." Translated by D. E. Linge. In *Philosophical Hermeneutics*. Edited by D. E. Linge. Berkeley: Univ. of California Press. 3-17.

Gadamer, H-G. 1994. *Truth and Method*. 2nd Revised ed. Translated by J. Weinsheimer & D. G. Marshall. New York: Continuum.

Gell-Mann, M. 1997. "Fundamental Sources of Unpredictability." *Complexity*, 3 (1): 9-13.

Gilb, T. 1995. *Software Inspection*. New York: Addison-Wesley.

Gilb, T. 1997. *Software Metrics*. Cambridge: Winthrop Publishers.

Gleick, J. 1987. *Chaos: Making a New Science*. New York: Penguin.

Goerner, S. 1995. "Chaos, Evolution, and Deep Ecology." In *Chaos Theory in Psychology and the Life Sciences*. Edited by R. Robertson and A. Combs. Mahwah, NJ: Lawrence Erlbaum Publishers. 17-38.

Grady, R., & D. Caswel. 1987. *Software Metrics*. New York: Prentice-Hall.

Grady, R. 1992. *Practical Software Metrics for Project Management and Process Improvement*. Englewood Cliffs, NJ: Prentice-Hall.

Guerlac, H. 1994. "Theological Voluntarism and Biological Analogies in Newton's Physical Thought." In *Philosophy, Religion, and Science in the 17th and 18th Centuries*. Edited by J. W. Yolton. Rochester, NY: Univ. of Rochester Press. 406-16.

Halverson, M., & D. Ozdes. 1992. "What Happened to the System Perspective in Reliability?" *Quality and Reliability Engineering International*, 8: 391-412.

Hammond, J., & J. R. Morrison. 1996. *The Stuff Americans Are Made Of*. New York: Macmillan.

Harvey, D. L., & M. Reed. 1997. "Social Science as the Study of Complex Systems." In *Chaos Theory in the Social Sciences*. Edited by L. D. Kiel and E. Elliot. Ann Arbor: Univ. of Michigan Press. 293-323.

Hawkins, M. 1997. *Hunting Down the Universe*. New York: Addison-Wesley.

Heady, R. B., & M. Smith. 1995. "An Empirical Study of the Topical Differences Between Total Quality Management and Quality Management." *Quality Management Journal*, 2 (3): 24-37.

Heelan, P. A. 1991. "Hermeneutical Phenomenology and the Philosophy of Science." In *Gadamer and Hermeneutics*. Edited by H. J. Silverman. New York: Routledge, Chapman and Hall. 213-228.

Heimann, P. M. 1994. "Voluntarism and Immanence: Conceptions of Nature in Eighteenth-Century Thought." In *Philosophy, Religion, and Science in the 17th and 18th Centuries*. Edited by J. W. Yolton. Rochester, NY: Univ. of Rochester Press. 393-405.

Holmes, N. 1998. "The Myth of the Computer Revolution." *IEEE Computer*, 31 (11): 121-22.

Horst, S. W. 1996. *Symbols, Computation, and Intentionality*. Berkeley: Univ. of California Press.

Humphrey, W. S. 1989. *Managing the Software Process*. New York: Addison-Wesley.

ISO. 1991. *ISO 9000-3 - Quality Management and Quality Assurance Standards – Part 3: Guidelines for the Application of ISO 9001 to the Development, Supply and Maintenance of Software.*

ISO. 1994. *ISO 9001 - Quality Systems - Model for Quality Assurance in Design, Development, Production, Installation and Servicing.*

Jacobson, I. 1992. *Object-Oriented Software Engineering*. Reading, MA: Addison-Wesley.

Jenner, M. G. 1995. *Software Quality Management and ISO 9001*. New York: John Wiley & Sons.

Jones, C. 1994. *Assessment and Control of Software Risks*. New York: Prentice-Hall.

Jones, C. 1997. "The Impact of Project Management Tools on Software Failures and Successes." *Proceedings of ProjectWorld 97*. Vol 1. Wash., DC: ProjectWorld Inc. C1/1-21.

Jones, K. G. 1991. *Messier's Nebulae and Star Clusters*. New York: Cambridge Univ. Press.

Juran, J. M., ed. 1988. *Juran's Quality Control Handbook*. 4th ed. New York: McGraw-Hill.

Kan, S. H. 1995. *Metrics and Models in Software Quality Engineering*. Reading, MA: Addison-Wesley.

Kaner, C. 1997. "Legal Issues Related to Software Quality." *Proceedings of the Seventh International Conference on Software Quality*, Montgomery AL, 6-7 October 1997. 2-13.

Kant, I. 1783. *Prolegomena to Any Future Metaphysics*. Carus translation edited by L. W. Beck. New York: Bobbs-Merrill.

Karama, K., et. al. 1991. "A Method for Software Reliability Analysis and Prediction Application to the TROPICO-R Switching System." *IEEE Transactions on Software Engineering*, 17: 334-344.

Kellert, S. H. 1993. *In the Wake of Chaos*. Chicago: Univ. of Chicago Press.

Kelsey, R. B. 1995a. "'A Plea for Tolerance in Matters Epistemological....'" *Software Engineering Notes*, 20: 38-39.

Kelsey, R. B. 1995b. "*PARSE: Problem Analysis and Process Management for Software Maintenance.*" 1995 Proceedings. *Pacific Northwest Software Quality Conference*, Portland OR, 27-28 September 1995. 423-33.

Kelsey, R. B. 1996. "Bad Fixes, Change Specifications, and Linguistic Constraints on Problem Diagnosis." *Software Engineering Notes*, 21: 74-78.

Kelsey, R. B. 1997. "Chaos Theory and Software Quality: Limitation or Invitation?" *Proceedings of the Seventh International Conference on Software Quality*, Montgomery AL, 6-7 October 1997. 179-187.

Kisiel, T. J. 1970. "Merleau-Ponty on Philosophy and Science." In *Phenomenology and the Natural Sciences*. Edited by J. J. Kockelmans & T. J. Kisiel. Evanston: Northwestern Univ. Press.

Kockelmans, J. J. 1966. *Phenomenology and Physical Science*. Pittsburgh: Duquesne Univ. Press.

Kockelmans, J. J., & T. J. Kisiel. 1970. Preface to *Phenomenology and the Natural Sciences*. Edited by J. J. Kockelmans & T. J. Kisiel. Evanston: Northwestern Univ. Press.

Kockelmans, J. J. 1984. *On the Truth of Being: Reflections on Heidegger's Later Philosophy*. Bloomington: Indiana Univ. Press.

Kockelmans, J. J. 1993. *Ideas for a Hermeneutic Phenomenology of the Natural Sciences*. Dordrecht, The Netherlands: Kluwer Academic Publishers.

Krensky, P. D. 1995. "ISO 9000: The Benefits & The Pitfalls." *1995 Conference Proceedings*. Nineteenth Annual Rocky Mountain Quality Conference, Denver CO, 12-13 July 1995. 204-210.

Lanubile, F. 1996. "Why Software Reliability Predictions Fail." *IEEE Software*, 13 (4): 131-137.

Lewenstein, B. V. 1997. "Life on Mars and in Science." Mercury, 26 (1): 24-26.

Loewer, B. 1998. "Copenhagen versus Bohmian Interpretations of Quantum Theory." Review of J. T. Cushing's *Quantum Mechanics: Historical Contingency and the Copenhagen Hegemony. British Journal for the Philosophy of Science*, 49: 317-28.

Lumley, T. 1997. "Complexity and the 'Learning Organization'." *Complexity*, 2 (5): 14-22.

Marick, B. 1995. *The Craft of Software Testing*. Englewood Cliffs, NJ: Prentice-Hall.

Martin, J., & C. L. McClure. 1983. *Software Maintenance: The Problem and Its Solutions*. New York: Prentice Hall.

McClure, C. L. 1981. *Managing Software Development and Maintenance*. New York: Van Nostrand Reinhold.

Merleau-Ponty, M. 1962. *Phenomenology of Perception*. Translated by C. Smith. New Jersey: The Humanities Press.

Merleau-Ponty, M. 1964a. "Einstein and the Crisis of Reason." In *Signs*. Translated by R. C. McCleary. Evanston: Northwestern Univ. Press. 192-197.

Merleau-Ponty, M. 1964b. "Indirect Language and the Voices of Silence." In *Signs*. Translated by R. C. McCleary. Evanston: Northwestern Univ. Press. 39-83.

Merleau-Ponty, M. 1964c. "On the Phenomenology of Language." In *Signs*. Translated by R. C. McCleary. Evanston: Northwestern Univ. Press. 84-97.

Merleau-Ponty, M. 1964d. "The Primacy of Perception and Its Philosophical Consequences." Translated by J. M. Edie. In *The Primacy of Perception*. Edited by J. M. Edie. Evanston: Northwestern Univ. Press. 12-42.

Milton, J. 1971. *The Complete Poetry of John Milton*. Edited by J. Shawcross. New York: Doubleday.

Möller, K. H., & D. J. Paulish. 1993. *Software Metrics*. London: Chapman & Hall.

Morrison, F. 1991. *The Art of Modeling Dynamic Systems*. New York: John Wiley & Sons.

Musser, G. S. 1997. "After the End of Science." *Mercury*, 26 (6): 22-32.

Nagel, E. 1979. *The Structure of Science: Problems in the Logic of Scientific Explanation*. Indianapolis: Hackett Publishing Co.

Nietzsche, F. 1966. *Beyond Good and Evil*. Translated by W. Kaufman. New York: Vintage.

Offut, A. J., & J. H. Hayes. 1996. "A Semantic Model of Program Faults." *Software Engineering Notes*, 21: 195-200.

Olson, D. 1993. *Exploiting Chaos: Cashing in on the Realities of Software Development*. New York: Van Nostrand Reinhold.

Oskarsson, Ö., & R. Glass. 1996. *An ISO 9000 Approach to Building Quality Software*. Upper Saddle River, NJ: Prentice-Hall.

Osler, M. J. 1994. "John Locke and the Changing Ideal of Scientific Knowledge." In *Philosophy, Religion, and Science in the 17th and 18th Centuries*. Edited by J. W. Yolton. Rochester, NY: Univ. of Rochester Press. 325-38.

Ott, E., & M. Spano. 1995. "Controlling Chaos." *Physics Today*, May: 34-40.

Parnas, D. 1998. "Software Engineering: An Unconsummated Marriage." *Conference Proceedings* of the Eleventh Annual Software Quality Week, Vol. 1, 26-29 May 1998, San Francisco, CA.

Paulk, M. C., et. al. 1995. Software Engineering Institute. *The Capability Maturity Model: Guidelines for Improving the Software Process*. New York: Addison-Wesley.

Peak, D., & M. Frame. 1994. *Chaos Under Control*. New York: W.H. Freeman & Co.

Peat, D. "Chaos: The Geometrization of Thought." In *Chaos Theory in Psychology and the Life Sciences*. Edited by R. Robertson and A. Combs. Mahwah, NJ: Lawrence Erlbaum Publishers. 359-72.

Penrose, R. 1996. *Shadows of the Mind*. New York: Oxford Univ. Press.

Perry, W. E. 1981. *Managing Software Maintenance*. Wellesley, MA: QED Information Sciences.

Perry, W. E. 1987. "Effective Methods of EDP Quality Assurance." In *Handbook of Software Quality Assurance*. Edited by G. G. Schulmeyer & J. I. McManus. New York: Van Nostrand Reinhold. 408-430.

Perry, W. E. 1995. *Effective Methods For Software Testing*. New York: John Wiley & Sons.

Peters, E. E. 1991. *Chaos and Order in the Capital Markets*. New York: John Wiley & Sons.

Pinch, T. 1997. "What's Cooking?" *Mercury*, 26: 29-31.

Plotinus. 1991. *The Enneads*. Translated by S. MacKenna. New York: Penguin.

Poulin, J. S. 1997. *Measuring Software Reuse*. New York: Addison-Wesley.

Pressman, R. S. 1997. *Software Engineering: A Practitioner's Approach*. 4th ed. New York: McGraw-Hill.

Prigogine, I. 1996. *The End of Certainty*. New York: The Free Press.

Pulford, K., A. Kuntzmann-Combelles, & S. Shirlaw. 1996. *A Quantitative Approach to Software Management*. New York: Addison-Wesley.

QAD. 1998. American Society for Quality Quality Audit Division. "Asia Pacific Organization Symposium on ISO-9000." *Vista*, 10 (3): 11-14.

QAI. 1997. *Solutions Manuals*. 4 vols. Orlando, FL: Quality Assurance Institute.

Raccoon, L. B. C. 1995a. "The Chaos Strategy." *Software Engineering Notes*, 20: 40-47.

Raccoon, L. B. C. 1995b. "The Chaos Model and the Chaos Lifecycle." *Software Engineering Notes*, 20: 55-66.

Radice, R. A. 1995. *ISO 9001 Interpreted For Software Organizations*. Andover, MA: Paradoxicon Publishing.

Rees, M. 1997. *Before the Beginning*. New York: Addison-Wesley.

Ricoeur, P. 1974. "Structure, Word, Event," Translated by R. Sweeney. In *The Conflict of Interpretations: Essays in Hermeneutics*. Edited by D. Ihde. Evanston: Northwestern Univ. Press. 79-96.

Ricoeur, P. 1976. *Interpretation Theory: Discourse and the Surplus of Meaning*. Fort Worth: The Texas Christian University Press.

Robertson, R. 1995. "Chaos Theory and the Relationship between Psychology and Science." In *Chaos Theory in Psychology and the Life Sciences*. Edited by R. Robertson and A. Combs. Mahwah, NJ: Lawrence Erlbaum Publishers. 3-15.

Ruelle, D. 1991. *Chance and Chaos*. New Jersey: Princeton Univ. Press.

Ruelle, D. 1997. "Chaos, Predictability, and Idealization in Physics." *Complexity*, 3 (1): 26-28.

Rumbaugh, J., et. al. 1991. *Object-Oriented Modeling and Design*. Englewood Cliffs, NJ: Prentice-Hall.

Sanders, J., & E. Curran. 1994. *Software Quality: A Framework for Success in Software Development and Support*. New York: Addison-Wesley.

Schmauch, C. H. 1994. *ISO 9000 for Software Developers*. Milwaukee: ASQC Quality Press.

Schrödinger, E. 1996. *Nature and the Greeks and Science and Humanism*. Canto edition. Cambridge: Cambridge Univ. Press.

Schulmeyer, G. G. 1997. "Software Quality Assurance Metrics." In *Handbook of Software Quality Assurance*. Edited by G. G. Schulmeyer & J. I. McManus. New York: Van Nostrand Reinhold. 318-342.

Seaman, C. B., & V. Basili. 1998. "Communication and Organization: An Empirical Study of Discussion in Inspection Meetings." *IEEE Transactions on Software Engineering*, 24 (7): 559-572.

Senge, P. M. 1990. *The Fifth Discipline*. New York: Doubleday.

Shakespeare, W. 1969. *William Shakespeare: The Complete Works*. New York: Viking.

Shaw, M. 1990. "Prospects for an Engineering Discipline of Software." *IEEE Software*, 7 (6): 15-24.

Shinbrot, T. 1993. "Chaos: Unpredictable yet Controllable?" *Nonlinear Science Today*, 3 (2):1-8.

Singh, M. P. 1992. *A Semantics for Speech Acts*. MCC Technical Report, CARNOT-138-92, Microelectronics and Computer Technology Corp., Austin (TX).

Sivak, J. M. 1998. "Adding Chaos to Improve Your Testing." *Conference Proceedings (CDROM) of the Seventh International Conference on Software Testing Analysis and Review*, Orlando FL, 4-8 May 1998. 1641-46.

Smith, P. 1998. "Approximate Truth and Dynamical Theories." *British Journal for the Philosophy of Science*, 49: 253-77.

Stapp, H. P. 1993. *Mind, Matter, and Quantum Mechanics*. New York: Springer-Verlag.

Szilasi, W. 1961. "Experience and Truth in the Natural Sciences." Translated by T. J. Kisiel. In *Phenomenology and the Natural Sciences*. 1970. Edited by J. J. Kockelmans & T. J. Kisiel. Evanston: Northwestern Univ. Press. 205-232.

Thalos, M. 1998. "A Modest Proposal for Interpreting Structural Explanations." *British Journal for the Philosophy of Science*, 49: 279-95.

TickIT. 1992. *A Guide to Software Quality Management System Construction and Certification Using EN29001*. London: DISC TickIT Office.

Torre, C. A. 1995. "Chaos, Creativity, and Innovation: Toward a Dynamical Model of Problem Solving." In *Chaos Theory in Psychology and the Life Sciences*. Edited by R. Robertson and A. Combs. Mahwah, NJ: Lawrence Erlbaum Publishers. 179-98.

Tsonis, P. A., & A. A. Tsonis. 1997. "Simplicity and Complexity in Gene Evolution." *Complexity*, 2 (5): 23-30.

Vandervert, L. R. 1993. "Neurological Positivism's Evolution of Mathematics." *The Journal of Mind and Behavior*, 14 (3): 277-288.

Vandervert, L. R. 1994. "How the Brain Gives Rise to Mathematics in Ontogeny and in Culture." *The Journal of Mind and Behavior*, 15 (4): 343-350.

Vincent, J., A. Waters, & J. Sinclair. 1988. *Software Quality Assurance*. New Jersey: Prentice-Hall.

Wang, Y., et. al. 1997a. "A Parallel Process Model for Software Quality Assurance." *Proceedings of the Seventh International Conference on Software Quality*, Montgomery AL, 6-7 October 1997. 54-66.

Wang, Y., et. al. 1997b. "Quantitative Evaluation of the SPICE, CMM, ISO 900 and BOOTSTRAP." *Proceedings of the 3rd IEEE International Symposium on Software Engineering Standards*. IEEE Computer Society Press. 57-68.

Weinstein, L. B., et. al. 1998. "What Higher Education Should Be Teaching About Quality – But Is Not." *Quality Progress*, 31 (4): 91-95.

Wheatley, M. J. 1992. *Leadership and the New Science*. San Francisco: Berrett-Kochler Publishers.

Whitrow, G. J. 1970. "Kant's Contributions to Cosmology and Cosmogony." In *Kant's Cosmogony*. Translated by W. Hastie. New York: Johnson Reprint Corp.

Williams, G. P. 1997. *Chaos Theory Tamed*. Wash., DC: Joseph Henry Press.

Wilson, P. F., L. D. Dell, & G. F. Anderson. 1993. *Root Cause Analysis: A Tool for Total Quality Management*. Milwaukee: ASQC Quality Press.

Yang, M. C. K., & A. Chao. 1995. "Reliability Estimation and Stopping Rules for Software Testing, Based on Repeated Appearances of Bugs." *IEEE Transactions on Reliability*, 44: 315-321.

Yeats, W. B. 1956. *The Collected Poems of W. B. Yeats*. New York: Macmillan Publishing Co.

Young, T. R. 1995. "Chaos Theory and Social Dynamics: Foundations of Postmodern Social Science." In *Chaos Theory in Psychology and the Life Sciences*. Edited by R. Robertson and A. Combs. Mahwah, NJ: Lawrence Erlbaum Publishers. 217-33.

Zuckerman, A. 1996. "European Standards Officials Push Reform of ISO 9000 and QS-9000 Registration." *Quality Progress*, 29 (9): 131-134.

INDEX

A

acceptance tester, 64
accrued cost, 61
Accuracy, 48, 49, 66

B

budget overrun, 70
bugs, 16, 129
business environments, 64, 65

C

Chaotic behavior, 4, 6, 7, 24, 50, 55, 59, 68, 73, 76, 80, 81, 82, 87, 90, 91, 92, 93, 94, 96, 97, 98, 99, 100, 101, 106, 135, 136, 142, 149
Chaotic transitions, 123
coarse grained analyses, 71, 74, 78
coarse grained strategy, 71, 76, 78, 82, 83, 87, 106, 112, 132, 152
code paths, 91, 93, 94, 98, 100, 101, 136, 143, 147
code production, 73
code reviews, 62, 79
Code structure, 63
coding, 15, 57, 66, 70, 71, 88, 125, 131, 135, 136, 140, 141, 146
coherence, 33, 49, 157
Communication protocols, 99
Correctness, 64, 74, 75, 76, 80

D

data exchange, 92
data integrity, 63, 141
deterministic relations, 39
development project, 14, 16, 58, 59, 60, 62, 67, 68, 71, 74, 86, 88, 90

E

elapsed time, 61, 147
Emergence, 42, 50
Error handling, 63
executable code, 62, 138
executed software, 90, 94
execution environment, 63

F

feedback, 96, 159
Flexibility, 75

H

hardware environment, 63, 94

I

information exchange, 10
initial conditions, 6, 34, 36, 42, 46, 54, 59, 73, 77, 86, 87, 88, 89, 90, 97, 107, 138, 147, 154, 159
Inputs, 63
Integrity, 74, 75

L

linguistic analysis, 73

M

macrolevel expectations, 92
Maintainability, 75
metaphor, 4, 42, 45, 54, 56, 82, 93, 157
microlevel expectations, 92, 93, 136
models, 1, 6, 11, 13, 15, 23, 35, 37, 38, 39, 41, 44, 45, 56, 71, 86, 87, 91, 105, 106, 107, 124, 134

N

nested systems, 62, 85, 87, 89, 148
networking, 10
nodes, 30, 92, 94, 99, 100, 127, 138, 140, 141, 143, 144, 145
nonlinearity, 93, 94, 101, 159

O

Operating system, 63
Operational complexity, 100
outputs, 63, 64, 74, 96

P

paradigm shift, 1, 2, 4, 28, 29, 30, 40, 41, 106, 107, 110, 134, 159
Portability, 4, 75

Processor speed, 64
product completion, 61
product quality, 2, 50, 86, 149
product specifications, 92, 131
productivity, 14, 60, 78, 79, 86, 105
project performance, 86
project review, 82

Q

quality assurance, 2, 5, 8, 10, 11, 16, 35, 45, 65, 148
queue management, 92, 98

R

Reliability, 64, 74, 75
reliability analysis, 45
retesting, 35, 79
Reusability, 75

S

scientific inquiry, 38, 111
scientific method, 8, 34, 41, 105, 133
social environment, 71
Software behavior, 93, 139
software development, 2, 4, 5, 8, 10, 11, 12, 13, 14, 15, 16, 17, 18, 22, 23, 24, 25, 35, 43, 47, 53, 54, 55, 56, 57, 58, 59, 68, 71, 76, 85, 86, 117, 125, 129, 148, 159
software industry, 2, 3, 4, 5, 6, 7, 9, 10, 13, 14, 16, 45, 46, 49, 56, 68, 86, 87, 89, 106, 107, 126, 132, 148, 149
software process, 17, 19, 20, 24
software professionals, 11, 12, 14, 19, 47, 48, 78, 87, 125
software quality, 3, 8, 12, 15, 35, 45, 47, 74, 77
software quality assurance, 3, 12, 15, 35, 45, 47
spheres of influence, 57, 58, 61, 62, 69, 70, 73, 82, 119, 135

statistical modeling, 39
Structural components, 63
structural strategy, 71, 78, 79, 82

T

Testability, 74, 75
top-down, 85, 86, 87
turbulence, 1, 6, 33, 44, 59, 77, 79, 82, 87, 88, 89, 90, 93, 94, 142, 143

U

Usability, 75, 77

W

worker productivity, 86
workflow efficiency, 78
Write Code Backward, 23
writing code, 4, 23, 61

Y

Year 2000 problem, 10